Key to the earthworms of Britain and Ireland

Second Edition

By Emma Sherlock

Field Studies Council

Publication code OP180.
ISBN 978 1 908819 40 6.

NATURAL HISTORY MUSEUM

The vision of the Natural History Museum, London is to advance our knowledge of the natural world, inspiring better care of our planet. The Angela Marmont Centre for UK Biodiversity at the Museum is for anyone with an interest in UK natural history.

To contact the Museum's Identification and Advisory Service, e-mail amc-enquiries@nhm.ac.uk, telephone 020 7942 5045, or join the online forums at www.nhm.ac.uk/natureplus.

Contents

Acknowledgements

All line drawings are by Louise Berridge. Photography by Harry Taylor and Phil Hurst, with greenhouse species taken by Debbie Flynn and Olha Schedrina. Additional photographs provided by Keiron Brown, Brad Griffiths, Samuel James and Olaf Schmidt.

Special thanks to Csaba Csuzdi, Keiron Brown and Rich Burkmar for reviewing the key both for its scientific accuracy and for setting up testing workshops to make sure the key is accessible to novice recorders. Thanks also to Frank Ashwood, Charlie Bell, Kerry Calloway and Bruce Sherlock for carefully reviewing the text and providing excellent feedback. Thanks to Simon Norman at Field Studies Council for guiding both the first and second editions to publication.

Like all AIDGAP keys, this guide was extensively tested during preparation and revised in the light of feedback from testers. This new edition has benefitted from the explosion of interest in earthworm recording that followed the publication of the First Edition in 2012. Today many more people are actively recording earthworms and submitting records to the national scheme. Building on the First Edition, successive drafts of the keys were tested and refined at workshops held throughout the UK in 2016 and 2017. Thanks to the participants at our many training courses, in particular:

Bernard Baverstock, Charlie Bell, Keiron Brown, Rich Burkmar, Kevin Clements, Ben Crabb, Heidi Cunningham, Julie Day, Sam Devine-Turner, Kate Fidczuk-Sterry, Frances Francis, Sasha Gorb, Harriet Green, Bob Griffiths, John Hatto, Sue Wend Ho, David Lewis, Rachelle Lewis, John A.W. Lock, Keith Lugg, Karen Lupton, Ana Morais Natalio, Martin Noble, Andrew O'Brien, Amy Orford, Anthony Roach, Natalie Sanders, Elwyn Sharps, Emma Sheard, Megan Shersby, Laura Sivess, Sue Townsend, Kate Tucker and Charles Turner.

Acknowledgments to First Edition (2012)

I would like to thank the Earthworm Society of Britain's committee members for helping with the testing of this key as well as anyone who has been on one of our many training courses.

Thank you to The Natural History Museum for their support, especially the Soil Biodiversity Group: Paul Eggleton, Daniel Carpenter, David Jones and Kelly Inward; and to the Zoology Department especially Miranda Lowe, Clare Valentine and Andrew Cabrinovic. A very big thank you to Louise Berridge for artistic support, and Phil Hurst and Harry Taylor for photography.

Thanks to Stewart Lee and Sally Mcphee for producing the datasheet used by the author and themselves on fieldwork in Nicaragua. Also thanks to Reginald Sims, my predecessor at the Natural History Museum and a huge inspiration. Without the key he developed with B.M. Gerard, I would never have broken into the world of British earthworms. Csaba Csuzdi, Chris Lowe, Kevin Butt, Aidan Keith, Gordon Paterson, Daniel Carpenter, Paul Eggleton and Miranda Lowe reviewed the text after it had been tested and made important suggestions for improvements.

Finally, thanks to Csaba Csuzdi and Victor Pop for answering my endless questions.

Introduction

Earthworms are some of the most commonly recognised soil invertebrates. They are found in urban parks, gardens, woodlands, farmland and even these days, with the introduction of worm bins, in our own homes. I think that most people appreciate the role that earthworms play in decomposition and healthy soils, but not many take the time to stop and study these highly important animals.

Surprisingly perhaps, the UK has a long history of earthworm scientists. Most are not household names, despite their great works, although one who is widely known is Charles Darwin. One of the most celebrated scientists to be born in the UK, he devoted years of his life to the study of earthworms and his last and in his lifetime biggest selling book, *The formation of vegetable mould through the action of earthworms*, published 1881, was entirely devoted to them.

Despite all of this, earthworms are still some of the least recorded animals on the planet. In the UK and Ireland we have only 31 species (including *Eisenia andrei*, a cryptic species of *E. fetida,* which is not keyed out separately here). Fortunately all but one of these species can be identified from external features alone. However we still have very few biological recorders for earthworms and consequently very little clue about their distribution and exact habitat preferences (Carpenter *et al.*, 2011). To get a full picture of what is happening around the UK in wet and dry soils, heaths, mires, and rotting logs we need to have many more recorders in many more parts of the country.

My main aim in producing this key is to encourage more people to record the earthworms throughout the UK and Ireland. While there have been good identification keys for British species in the past (and up until 2002 they accounted for all the Irish earthworms as well) I have realised when taking courses that people still struggle to get to grips with the main features. I hope this key, which is illustrated with photographs at every stage, will help.

I also wanted to put together for the first time in one key all the earthworms that are likely to be found naturally in the soils in the UK and Ireland regardless of family, and update the taxonomy which has changed since the last keys were produced.

Scope of the key

This key is for the identification to species level of mature preserved specimens of all species of earthworm in the UK and Ireland. The species listed are those that might be found living freely in the soils around the UK (Table 1 in Sherlock & Carpenter, 2009, with corrected omissions). The worms that are only found in hot greenhouses around the country, but have so far been unable to colonise the soils outside of these environments, are included in the list but excluded from the main key. In this 2nd edition a key to those species which seem to be a more permanent feature of greenhouses is also included. Taxonomic names are updated from Csuzdi & Zicsi (2003), Blakemore (2008) and Csuzdi *et al.* (2017).

Preserved worms are more likely to be accurately identified than live specimens. A field key to live specimens of a small number of common species was produced as part of the Open Air Laboratories (OPAL) project in 2009. However, for accurate identifications for all species, preservation is required (OPAL, 2011).

Systematic list of the earthworms of the UK and Ireland

Table 1 contains a complete list of earthworm species recorded in the UK and Ireland. Part 1 lists the species found living in natural environments.

Part 1: earthworm species living in natural environments				
Family	Genus	Species	Authority	Notes
Acanthodrilidae	*Microscolex*	*phosphoreus*	(Duges, 1837)	
Lumbricidae	*Allolobophora*	*chlorotica*	(Savigny, 1826)	Two colour morphs: pink and green
Lumbricidae	*Aporrectodea*	*caliginosa*	(Savigny, 1826)	
Lumbricidae	*Aporrectodea*	*cupulifera*	(Tetry, 1937)	Found in Ireland 2002
Lumbricidae	*Aporrectodea*	*icterica*	(Savigny, 1826)	
Lumbricidae	*Apporectodea*	*nocturna*	(Evans, 1946)	Formerly considered a morph of *A. caliginosa*
Lumbricidae	*Aporrectodea*	*limicola*	(Michaelsen, 1890)	
Lumbricidae	*Aporrectodea*	*longa*	(Ude, 1885)	
Lumbricidae	*Aporrectodea*	*rosea*	(Savigny, 1826)	
Lumbricidae	*Bimastos*	*eiseni*	(Levinsen, 1884)	Called *Lumbricus eiseni* by Sims & Gerard 1999 but *Allolobophoridella* by Blakemore 2005 and Csuzdi & Zicsi 2003; changed to *Bimastos* by Csuzdi *et al.* 2017
Lumbricidae	*Bimastos*	*rubidus*	(Savigny, 1826)	Formerly *Dendrodrilus* changed to *Bimastos* by Csuzdi *et al.* 2017
Lumbricidae	*Dendrobaena*	*attemsi*	(Michaelsen, 1902)	
Lumbricidae	*Dendrobaena*	*hortensis*	(Michaelsen, 1890)	Called *Eisenia hortensis* by Sims & Gerard 1999 but *Dendrobaena* by Blakemore 2005 and Csuzdi & Zicsi 2003
Lumbricidae	*Dendrobaena*	*octaedra*	(Savigny, 1826)	
Lumbricidae	*Dendrobaena*	*pygmaea*	(Savigny, 1826)	
Lumbricidae	*Dendrobaena*	*veneta*	(Rosa, 1886)	Called *Eisenia veneta* by Sims & Gerard 1999 but *Dendrobaena* by Blakemore 2005 and Csuzdi & Zicsi 2003
Lumbricidae	*Eisenia*	*fetida*	(Savigny, 1826)	
Lumbricidae	*Eiseniella*	*tetraedra*	(Savigny, 1826)	
Lumbricidae	*Helodrilus*	*oculatus*	Hoffmeister, 1845	

Family	Genus	Species	Authority	Notes
Lumbricidae	*Keenleenus*	*armadus*	Blakemore, 2012	Discovered in Ireland
Lumbricidae	*Lumbricus*	*castaneus*	(Savigny, 1826)	
Lumbricidae	*Lumbricus*	*festivus*	(Savigny, 1826)	
Lumbricidae	*Lumbricus*	*friendi*	Cognetti, 1904	
Lumbricidae	*Lumbricus*	*rubellus*	Hoffmeister, 1845	
Lumbricidae	*Lumbricus*	*terrestris*	Linnaeus, 1758	
Lumbricidae	*Murchieona*	*muldali*	(Omodeo, 1956)	Named *miniscula* by Sims & Gerard 1999, but later that year *muldali* (an earlier synonym of *miniscula*) was resurrected as a separate species; *miniscula* is a southern species, *muldali* a northern one
Lumbricidae	*Octolasion*	*cyaneum*	(Savigny, 1826)	
Lumbricidae	*Octolasion*	*lacteum*	(Örley, 1881)	Called *tyrtaeum* by Sims & Gerard 1999 but this was based on what is widely regarded as a false synonomy made by Gates
Lumbricidae	*Satchellius*	*mammalis*	(Savigny, 1826)	
Sparganophilidae	*Sparganophilus*	*tamesis*	Benham, 1892	

Part 2: earthworm species recorded from artificial envionments

Family	Genus	Species	Notes	Most recent records
Acanthodrilidae	*Dichogaster*	*affinis*		source unclear?
Acanthodrilidae	*Dichogaster*	*bolaui*	Friend recorded a new species *lageniformis* but Sims believed the form to be probably *boulai*	1916, Kew (Friend)
Acanthodrilidae	*Dichogaster*	*saliens*		1893, Kew (Beddard)
Eudrilidae	*Eudrilus*	*eugeniae*		1906, Kew (Beddard)
Eudrilidae	*Eudriloides*	*durbanenesis*		1893, Kew (Beddard)
Eudrilidae	*Heliodrilus*	*lagosensis*		1893 (Beddard)
Eudrilidae	*Hyperiodrilus*	*africanus*		1893 (Beddard)
Eudrilidae	*Iridodrilus*	*roseus*		
Megascolecidae	*Amynthas*	*alexandri alexandri*		
Megascolecidae	*Amynthas*	corticis	Often parthenogenetic	2015, Edinburgh (Lugg). 2016, Kew

Continued

Family	Genus	Species	Notes	Most recent records
Megascolecidae	*Amynthas*	*gracilis*	Parthenogenic morph of *A. corticis*	2016, Kew. 2016, Wisley (Lugg)
Megascolecidae	*Amynthas*	*morrisi*		
Megascolecidae	*Amynthas*	*rodericensis*		2007, 2012 & 2016, Kew (Sherlock). 2015, Edinburgh (Lugg)
Megascolecidae	*Anisochaeta*	*celmisiae*		1975 (Jamieson); but not established?
Megascolecidae	*Anisochaeta*	*minor*		1975; thought by Sims to have established?
Megascolecidae	*Duplodicodrilus*	*californica*		2012, Kew
Megascolecidae	*Metaphire*	*posthuma*	Misidentified originally by Beddard as *Perichaeta sumatra* (Sims 1999)	1906 (Beddard)
Megascolecidae	*Metaphire*	*schmardae schmardae*	Misidentified originally by Beddard as *Perichaeta sumatra* (Sims 1999)	1892 (Beddard)
Megascolecidae	*Polypheretima*	*taprobanae*		2009, Kew
Megascolecidae	*Pithemera*	*bictincta*		2007, 2012 & 2016, Kew (Sherlock). 2012, Dublin (Schmidt)
Megascolecidae	*Perionyx*	*exavatus*		1986 (Easton)?
Moniligastridae	*Drawida*	*barwelli*		1893, Kew (Beddard)
Ocnerodrilidae	*Eukerria*	*rubra*		1916, Oxford (Friend)
Ocnerodrilidae	*Gordiodrilus*	*ditheca*		1893, Kew (Beddard)
Ocnerodrilidae	*Gordiodrilus*	*dominicensis*		1893, Kew (Beddard)
Ocnerodrilidae	*Gordiodrilus*	*elegans*		1893, Kew (Beddard)
Ocnerodrilidae	*Gordiodrilus*	*robustus*		1893, Kew (Beddard)
Ocnerodrilidae	*Gordiodrilus*	*tenuis*		1893, Kew (Beddard)
Ocnerodrilidae	*Nannodrilus*	*africanus*	Originally called *Ilyogenia*	1893, Kew (Beddard)
Ocnerodrilidae	*Nematogenia*	*lacuum*	Originally called *Ocnerodrilus*	1893, Kew (Beddard)
Ocnerodrilidae	*Ocnerodrilus*	*occidentalis*		source unclear?
Octochaetidae	*Eutyphoeus*	*nicholsoni*		1901, Kew (Beddard)
Octochaetidae	*Trigaster*	*minima*		1911, Kew (Friend)
Rhinodrilidae	*Diachaeta*	*hesperidum*		
Rhinodrilidae	*Diachaeta*	*barbadensis*		
Rhinodrilidae	*Pontoscolex*	*corethrurus*		2007, 2012 & 2016, Kew (Sherlock)

Ecology

Importance of earthworms

The importance of earthworms to soil health and structure has long been recognised. Aristotle described them as the 'intestines of the earth' which is very apt due to the way they break down dead organic matter, processing it through their gut with the aid of symbiotic bacteria and a crop and gizzard system, and release nutrients back into the soil.

More recently they have been described as 'ecosystem engineers' (Edwards & Bohlen, 1996; Lavelle *et al.*, 1997). This is due to the fact that they directly influence and change the environment around themselves and the resources available for other organisms (Butt & Grigoropoulou, 2010).

Figure 1. Earthworm casts.

1. Earthworm burrowing (see p. 6) improves soil aeration and drainage. Earthworms are crucial for ameliorating the effects of intense rainfall, digging more burrows in these conditions, thus allowing water to dissipate more rapidly (Andruizzi *et al.*, 2015).
2. Soil gut passage enhances soil crumb structure and nutrient availability for other organisms alongside bacteria and fungi which increase in number with earthworm presence.
3. The process of pulling dead organic matter down into burrows increases nutrient availability, especially lower in the soil profile, and improves the water holding capacity of the soil.

Earthworms can also play a part in the mineral weathering of rock. *Dendrobaena veneta* has been proven to accelerate the weathering of amorthite, biolite, smeetite and kaolinite (Carpenter *et al.*, 2007), and in the Carpathian Mountains *Octodrilus* species have affected the mineralogy and formation of illite soils (Pop, 1998). The work of earthworms can often determine the depth that roots can penetrate, the amount of water that can be stored, and the movement of air within the soil (Hermavan & Cameron, 1993).

Soil organisms like earthworms improve the resilience of soil against disturbance (Brussaard *et al.*, 2007), and coupled with agricultural practices that stimulate soil biodiversity, such as increased crop diversity, reduced tillage and continuous cover, could help mitigate the effects of climate change (de Vries *et al.*, 2012). They really are little engineers.

Numbers of earthworms in the soil and what they do

It is difficult to start putting figures on what earthworms do for us. A report from Ireland attempted to estimate the financial services earthworms provide, reporting it to be worth in the region of 700 million euros per year. (Blakemore, 2010). However very few studies have put figures on exactly the work earthworms are doing and it varies widely from area to area.

Soils in the UK and Ireland usually contain anywhere from 10 to 400 individual worms per square metre. Darwin (1881) estimated 53 000 worms per acre (approximately 13 per square metre, a conservative estimate), while Friend (1924) estimated 160 000 per acre (approximately 40 per square metre). Lee (1985) sums up that in Northern European soils numbers of worms can range from under 10 to over 400 per square metre, and occasionally in forests or highly fertile pastures 500 to 1000 per square metre.

Finding figures on worm movement and work is even trickier. *Lumbricus terrestris* itself (so just 1 species) is thought to move 1.2 tonnes of dead leaves from a hectare of orchard per year (90% of autumn leaf fall) (Edwards & Lofty, 1977). Earthworms are thought to cast 27 tonnes of soil per hectare in a pasture (Guild, 1955) but not all environments contain the same number of worms, so this will vary from site to site.

Habitats and burrowing systems

Earthworms can be found in most habitats in the UK and Ireland, except those that are too dry or too toxic. They are rarely found in very acidic soils. Some species can go into diapause to survive the driest (or coldest) parts of the year. They burrow much deeper, knot their body into a ball and envelope themselves in mucus. This reduces the surface area and creates a humid microclimate around the body. Diapause can last for several months in tropical species and is vital for their survival.

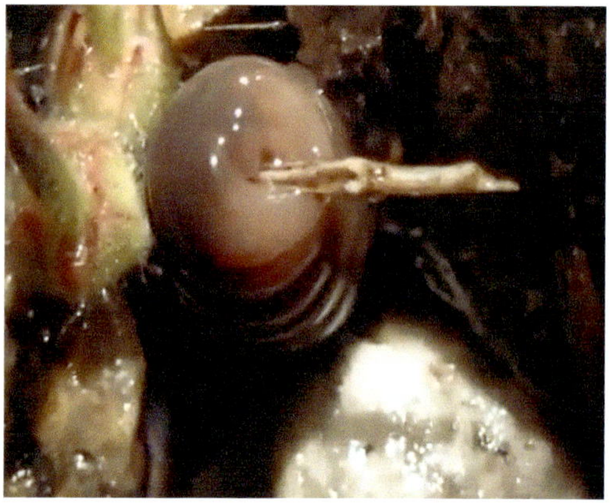

Figure 2. Earthworm burrowing behaviour (Brad Griffiths).

With earthworms come a network of burrows: some vertical, up to 2.5 metres deep (Edwards & Bohlen, 1996) and almost unbranched (*Lumbricus terrestris*, *L. friendi* and *Aporrectodea nocturna*) (Nuutinen & Butt 2003); some vertical but less deep with branches (*Aporrectodea longa*) (Andriuzzi *et al.*, 2016); and some horizontal (endogeic worms). *Allolobophora chlorotica*, a typical endogeic worm, lives in transient burrows near the soil surface, ingesting soil to feed on organic matter and living organisms (Andriuzzi *et al.*, 2016).

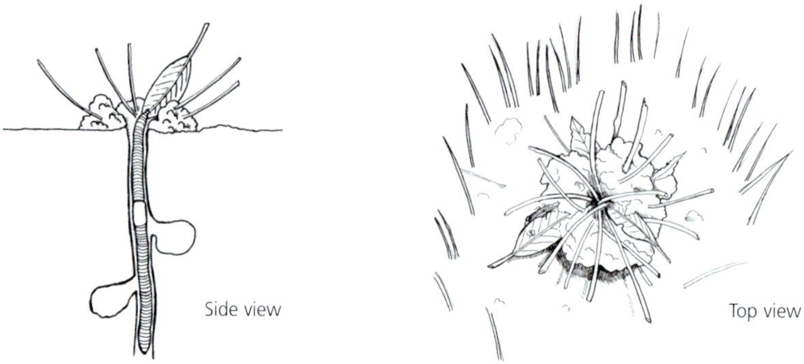

Figure 3. *Lumbricus terrestris* and *L. friendi* create middens at the entrance to their burrows: little piles of casts and organic matter on the soil surface. The mean distance between middens is 13 cm (Grigoropoulou & Butt, 2010).

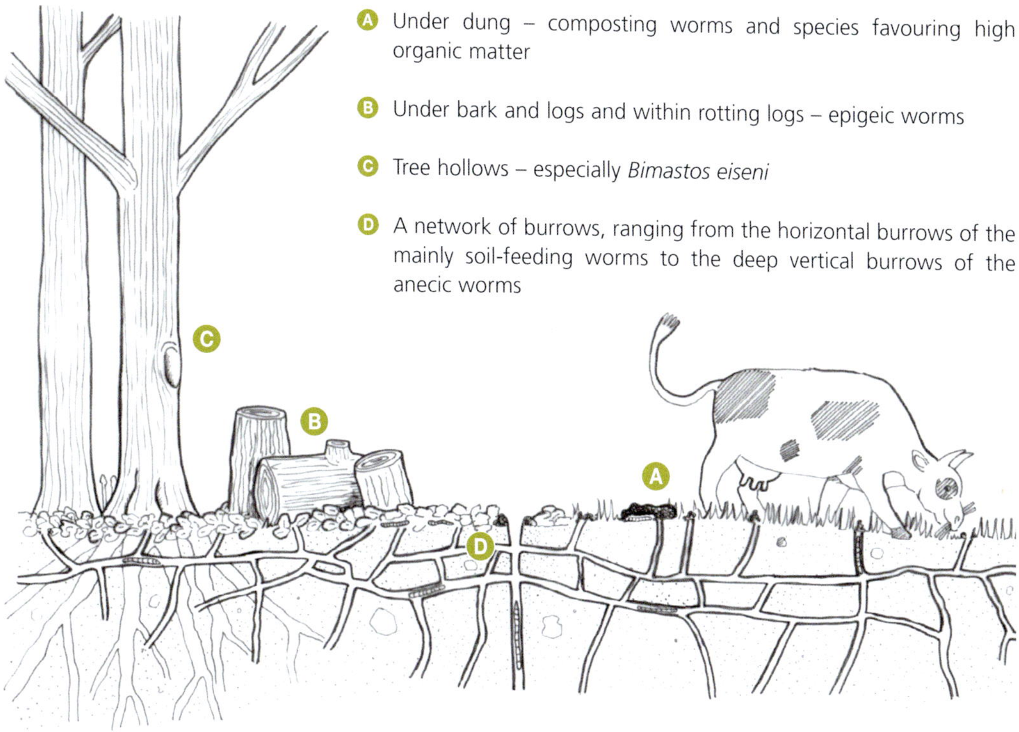

Figure 4. Where to find earthworms in a wooded or rural area (Louise Berridge).

Ecological groupings

As this guide only covers UK and Ireland species, the ecological earthworm grouping models of Blakemore (2008) or Lee (1959) have not been used; instead the four ecological groups here are based on Bouche (1971, 1977).

Group 1: the composters

These are mainly found in compost heaps or in areas of very high organic matter such as under cow pats in a field. Usually no soil is consumed. They are red in colour and often have a stripy appearance when stretched out.

(This group can be lumped with the epigeic worms, but these particular species need exceptionally high organic content so are often found separately.)

Figure 5. *Eisenia* spp. in a compost bin.

UK and Ireland species: *Eisenia fetida*, *Eisenia andrei*, *Dendrobaena hortensis*, *Dendrobaena veneta*.

Group 2: the epigeics

These are surface-dwelling earthworms found under leaf litter and in rotting logs. They feed directly on the decaying matter. Little or no soil is consumed. They are usually red in colour.

Figure 6. *Lumbricus rubellus*: an epigeic worm.

UK and Ireland species: *Dendrobaena attemsi*, *Dendrobaena octaedra*, *Dendrobaena pygmaea*, *Bimastos eiseni*, *Bimastos rubidus*, *Lumbricus castaneus*, *Lumbricus festivus*, *Lumbricus rubellus*, *Satchellius mammalis*.

Group 3: the endogeics

These are the soil dwellers with horizontal burrows. They rarely come to the surface or consume the soil itself. Often grey, green or pink in colour and slightly pigmented.

Figure 7. *Apporectodea rosea*: an endogeic worm.

UK and Ireland species: *Allolobophora chlorotica*, *Aporrectodea caliginosa*, *Aporrectodea cupulifera*, *Aporrectodea icterica*, *Aporrectodea limicola*, *Aporrectodea rosea*, *Eiseniella tetraedra*, *Helodrilus oculatus*, *Murchieona muldali*, *Octolasion cyaneum*, *Octolasion lacteum*.

Group 4: the anecics

These are the deep burrowing worms with vertical burrows. *Lumbricus terrestris* in particular comes up to the surface at night and draws leaves and other dead plant material down into its burrow. Some soil can be ingested and for *Aporrectodea longa* more soil is ingested than *L. terrestris*. They are often a deep red or even black colour dorsally and are always large worms.

Figure 8. *Apporectodea longa*: an anecic worm.

UK and Ireland species: *Aporrectodea nocturna*, *Aporrectodea longa*, *Lumbricus friendi*, *Lumbricus terrestris*.

In reality earthworms do not conform rigidly to these groups. Adult *Lumbricus rubellus* can, for example, be as little as 1.5 cm long as well as 20 cm long, and be found at 50 cm deep in the soil as well as at the soil surface. Different individuals might be grouped as epigeics or anecics.

Further reading on earthworm ecology: Blakemore (2008); Edwards (2004); Edwards & Bohlen (1996); Lee (1985).

Biology

Anatomy

Earthworms are in the phylum Annelida, class Clitellata and subclass Oligochaeta.

Class Clitellata consists of the two subclasses Hirudinea (leeches) and Oligochaeta. Both are segmented worms which possess a clitellum (or saddle). Leeches are the closest living relatives of earthworms (see Fig. 9, the leech *Trochaeta subviridis* with a recognisable clitellum and segmentation).

However earthworms have very different feeding structures and a number of different internal structures too.

Figure 9. *Trocheta subviridis*, a common terrestrial leech found deep in clay soils throughout the UK and Ireland.

To understand how earthworms function, some of their internal anatomy, which is readily seen on dissection, is included here.

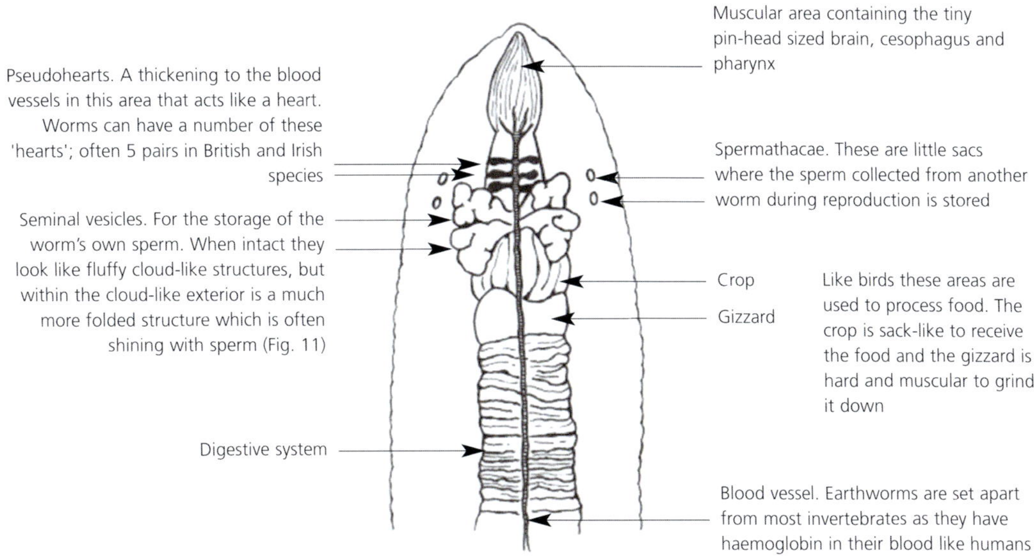

Figure 10. Internal features of an adult earthworm.

The digestive system of an earthworm involves a long gut through the body. Near the start of the gut is a soft crop and hard muscular gizzard below. Within the gizzard, organic matter like dead leaves is physically broken down by grinding. Crushed egg shells can be a useful source of grinding material in compost heaps.

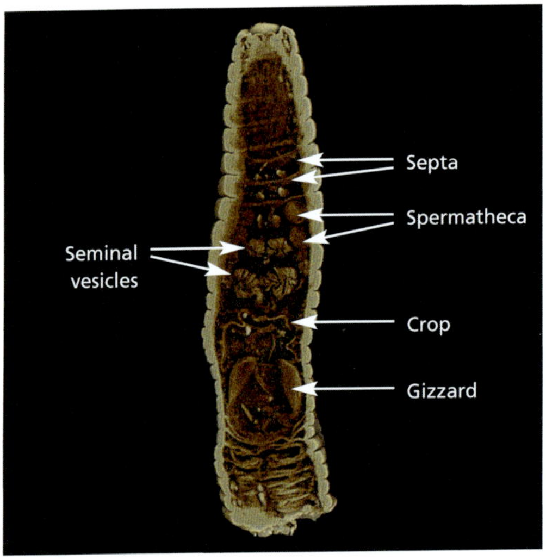

Figure 11. CT scan of an earthworm (Hall, Sherlock & Sykes, 2015).

Leading from the gizzard, the soft gut is full of symbiotic bacteria which assist in the further breakdown of the organic matter. The worms possess califerous glands, which vary in form, position and size in different species, and can be difficult to see. Their function is to secrete calcium carbonate as a milky fluid into the oesophagus; worms living in acidic environments and consuming fresh organic matter tend to have more complex glands with a greater amount of secretions (Briones *et al.*, 2008).

Regeneration

Earthworms have the ability to regenerate parts of their bodies. If some species of earthworm are disturbed they can start to detach their tail end.

Figure 12. *Apporectodea longa* starting to detach its tail.

This can be an effective escape mechanism: the tail will keep twitching after detachment for a few minutes and may distract predators while the worm itself escapes to safety. For any gardeners that accidently slice a tail off a worm, it will regenerate a new one – so they don't need to feel too guilty!

If, however, the head end is damaged, the worm is less likely to survive, although a worm can eventually regenerate its 'brain'. Losing these first twenty or so segments which contain most of the vital organs would be tough to regenerate. When losing a tail the worm leaves itself open to infection. Although they lack real antibodies, their skin secretions do contain antibacterial peptides (Wang, 2005).

'The underlying mechanisms which drive or repress complete functional and structural neuronal regeneration have been elusive. In our efforts to understand this complex phenomenon, the earthworm can be utilised as a model animal, since it can fully regenerate its cerebral ganglion (loosely defined as the 'brain') following its precise surgical removal. The characterisation of fundamental aspects of neuronal regeneration in the earthworm promises to provide an insight as to why humans have largely lost that capability.

'In Charles Darwin's last book, *Earthworms*, published in 1881, he acknowledged the importance of earthworms in nature – perhaps he was right in a much bigger way than he could have ever imagined.' Antreas Katsiamides (pers. comm.) www.toxicogenomics.info

Locomotion

Figure 13. *Lumbricus terrestris* with its flattened tail.

The main anatomical features for locomotion are two opposing muscle groups (circular muscles and longitudinal muscles) all along their body and a fluid filled cavity (the coelom). The coelom gives the body the rigidity it needs for movement. It is the expansion and contraction of these muscles working against each other and against the rigid coelom that moves the body along. When the circular muscles (on the outside) contract, the worm becomes long and thin. When the longitudinal muscles (underneath the circular muscles) contract, the worm becomes shorter. Using its setae to grip, the worm can then move forward by this shortening and lengthening process.

Some anecic worms that forage or mate on the surface use the tail as an anchor. Most of the body emerges, but the tail stays within the burrow, ready to pull the worm straight back down if danger approaches. The setae, or small hairs along the body, help with digging and gripping the soil. Some species have a distinctively flattened tail to help this process.

Reproduction

Earthworms are hermaphrodites, with both male and female structures. Most species reproduce sexually, as cross-fertilising hermaphrodites, in a process which involves two adult worms exchanging sperm with each other. In *Lumbricus terrestris* this often happens above ground.

Figure 14. *L. terrestris*: the male pore of one worm is aligned to the spermathecal pores of the other (Brad Griffiths).

Figure 15. *L. terrestris*: the two worms seal themselves together with mucus (Brad Griffiths).

The worms arrange themselves so the spermathecal pores of one worm is aligned with the clitellar region of the other. The tubercula pubertatis (TP) produces mucus to seal the two animals together. Sperm is released from the seminal vesicles of each worm and leaves the body via the male pores. The sperm travels along the first worm's body (some species have special grooves to facilitate this) and is then taken in by the spermathecal pores of the opposite worm and stored in little sacks called spermathecae. These two male systems (seminal vesicles and spermathecae) within the worm are not interconnected (see Fig. 17, page 12).

Later the clitellum produces a thick band of mucus, which travels over the head end of the worm as the worm pulls itself backward out of it. When the head passes the female pores, the worm's own egg or eggs (depending on the species) are then released. When the band passes the spermathecal pores, the stored sperm from the other worm is also released into it. Once the mucus has passed over the head, it hardens and forms a little cocoon in the soil (Fig. 16).

Figure 16. Cocoon from a compost heap.

Different species vary in the number of juvenile earthworms that emerge from the cocoon, how long emergence takes, and the time needed for juveniles to mature. Environmental conditions can also play a role. See the species accounts (pages 42-72) for information we currently have.

Mostly this is an opportunistic activity, but when partners are plentiful some selection appears to take place. Larger worms of a species will generally select another larger individual for mating, as shown by experiments with *Eisenia fetida* and *Lumbricus terrestris*. This is thought to be due to heavier worms producing denser cocoons and larger offspring, not more individuals (Cosín *et al.*, 2011). It has even been documented for *Lumbricus terrestris* that an individual will insert its anterior few segments into the potential mate's burrow to assess the quality (Nuutinen & Butt, 1997).

Self-fertilisation is unusual, but not unknown. *Eisenia andrei* (paired in this key with *Eisenia fetida*) twists its body to enable its spermathecal pores to contact the ventral surface of its clitellum (Cosín *et al.*, 2011). Sperm is then transported via the male pores to the spermathecae.

The other reproductive strategy is parthenogenesis; in short, producing clones. No sperm is involved so often worms using this mode of reproduction will have reduced male structures and there will be no shining or irredescence of the male structures internally. There are no spermatophores cropping up on the earthworm's surface. Parthenogenesis is thought just to occur within two families: Lumbricidae and Megascolecidae (Cosín *et al.*, 2011).

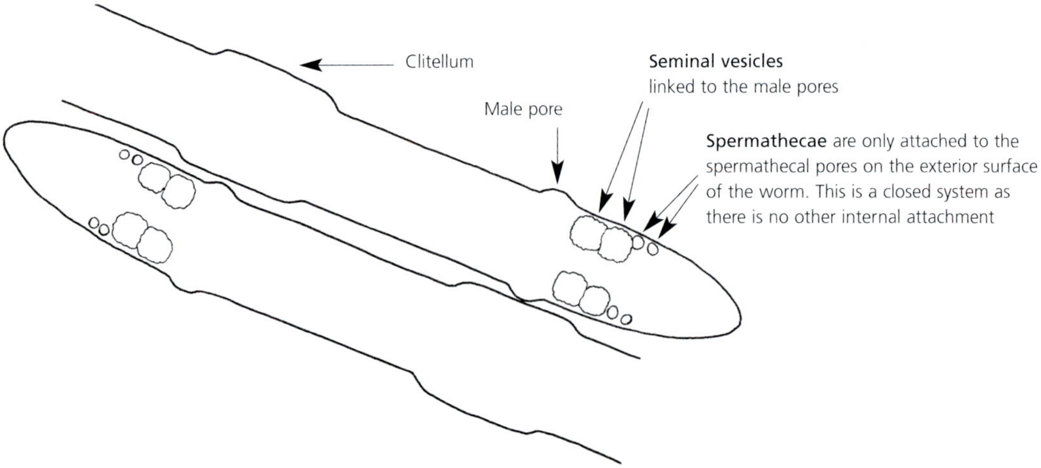

Figure 17. Stylised representation of earthworm mating.

Spermatophores are small capsules that adhere to the body wall (often seen when identifying worms). Edwards & Bohlen (1996) suggest that in some cases sperm does not reach the spermathecae but instead forms a small capsule on the body surface. As the mucus, which will become the coating of the cocoon, makes its way along the body the spermatophore gets incorporated directly into the cocoon.

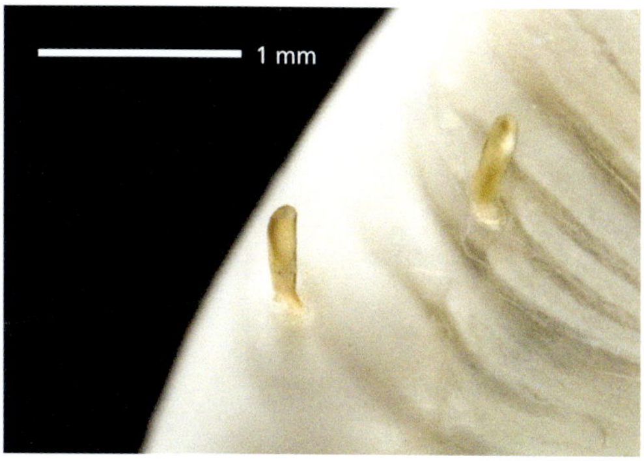

Figure 18. Spermatophore.

Predation

Earthworms are a crucial part of the food web, as prey for many animals. Birds are major predators; it is estimated that earthworms make up 90% of the food source of black-headed gulls in autumn (Cuendet, 1983). Moles, badgers, foxes and shrews are the largest consumers of earthworms among the mammals.

It has been estimated that male moles consume 18-36 kg of earthworms annually in Britain, accounting for 50-100% of their diet depending on the season (Mellanby, 1966). This is not a straightforward process. Mole saliva contains a toxin which paralyses the earthworms. They can then be stored in 'larders' underground to be consumed when needed. There can be up to 1000 earthworms in any one 'larder'. Badgers are more picky, predominantly predating on *Lumbricus terrestris*, although other species can be consumed. An adult is estimated to eat 130-200 individuals each per night. (Kruuk, 1978). Foxes can eat ten worms a minute when foraging (Macdonald, 1983) and for the cubs their survival at certain stages of development can be reliant on a good source of earthworms. At certain times of the year, shrews need earthworms for up to 60% of their food.

Figure 19. Beetle larva feeding (Brad Griffiths).

Some invertebrates prey on earthworms as well. These include flatworms, (both introduced and native), leeches (especially terrestrial leeches), beetle larvae and fly larvae. For example the cluster fly lays eggs in the earthworm, which later becomes the first meal of the larvae on hatching.

With so many predators, do earthworms have any defences against predators? As noted above (p. 10) some species can drop a tail, with the hope the predator will go for the twitching tail rather than the worm. Another defence mechanism that *Eisenia fetida* in particular can use is to secrete a bright yellow coelomic fluid. This fluid has a distinctive smell and is bright yellow, a colour often associated with poison. Although not toxic it is likely that it can deter birds. (Fig. 20).

Figure 20. Yellow fluid produced by *Eisenia fetida*.

The global picture

Over 6000 species in 18 different families have been discovered, occurring on all continents except Antarctica. They are found in the same range of habitats as in the UK and Ireland, with the addition of life within epiphytes in the tree canopy (often associated with the root masses and between fronds), and, in one case, a salt-tolerant species living in marine littoral zones (*Pontodrilus litoralis*).

Many earthworm species are cosmopolitan or peregrine species that have spread with humans and now occur widely. *Pontoscolex corethurus* is one of the most well known. However our own *Lumbricus terrestris*, *L. caliginosa* and *Eisenia fetida* have also travelled and settled.

Large areas of the world are completely unexplored for their earthworm fauna. With deforestation and the threat of climate change, it is possible that many species are disappearing before they are even found.

Figure 21. *Eutrigaster (Graffia) azul*, Nicaragua (Emma Sherlock & Csaba Czudi).

Figure 22. *Archipheretima middletoni*, Philippines (Samuel James).

Earthworms are very diverse in colour and size. Within the UK and Ireland they can be green, pink or blueish, have black heads and deep red heads, and grow to an adult size ranging from 1-40 cm. There are a number of bright blue worms from Australia, and central and south America (Fig. 21). Even stranger is a fantastic discovery from The Philippines by Samuel James in 2011 of *Archipheretima middletoni*, the fried egg worm, the only earthworm known to contain a poison. So they are not only the one colour that common perceptions would have you believe.

No known fossil earthworms exist (Anderson *et al.*, 2017). However, fossil evidence of earthworm-like organisms has been found, such as burrows, faecal material and other fossilised structures, but since earthworms are soft-bodied animals they do not preserve properly in the fossil record. Leech cocoon fossils have been discovered from the late Triassic (approximately 200 million years ago). As leeches are the closest living relatives of the oligochaetes, it can be inferred that the minimum age of the common ancestor of earthworms and leeches must be around this time or earlier. Anderson *et al.* (2017) suggest that the ancestor of all living earthworms probably lived over 209 million years ago.

Table 2. Summary of earthworm families.

Family	Distribution	Notes
Acanthodrilidae	Mainly South America	Includes Benhaminae
Ailmidae	Mainly Africa and Asia	Mainly aquatic
Ailoscolecidae	North America and high altitudes in Europe	
Biwadriloidae	Japan only	
Eudrilidae	Africa	Often organs displaced to much further down the body
Glossoscolecidae	Mainly South America, but also found in the Caribbean	Mostly forest dwellers
Hormogastridae	Mediterranean Europe	
Kazimierzidae	South Africa	Unique circulatory system
Kynotidae	Madagascar	
Lumbricidae	Mainly European, with a few endemic species in the Far East. Some cosmopolitan earthworms have spread globally	Clitellum furthest back of all the earthworm families, generally starting after segment 20
Lutodrilidae	North America	Aquatic
Megascolecidae	Mostly Asia and Australia. Some cosmopolitan species are global, and have been found in greenhouses in the UK	Clitellum generally 14-16; very active worms; often with very diverse male and prosthetic pores
Microchaetidae	South Africa only	Spermathecae tiny attached to the body wall; some worms very large
Monilgastridae	Far East and India	
Ocnerodrilidae	Mainly Africa, South America and India	Often aquatic
Octocheatidae	Mainly South America, but also Australasia, India, the Far East and Africa	
Rhinodrilidae	South America, although the cosmopolitan *Pontoscolex corethurus* is found globally and often considered a pest	
Sparganophilidae	Mainly North American family. One species found in the UK	Aquatic
Tritogeniidae	South Africa	
Tumakidae	South America	

Sampling techniques

A number of different sampling methods are required to get a comprehensive picture of all the different species of earthworm to be found in a particular habitat.

The soil

When digging for earthworms the best method is to dig out blocks of soil. First mark out a square of 25 x 25 cm with a spade (in practice about a spade and a half by a spade and a half). Consistency in sampling is essential so that different sites can be compared at different dates; particularly important if any ecological studies are to be undertaken.

Once the square has been marked, as quickly as possible dig the block out, preferably whole. Depending on the soil and density of roots and stones this may not always be possible. Aim for a depth of 20 cm if possible.

Figure 23 (right). Mark a 25 x 25 cm square with a spade, then dig out a block of soil.

Figure 24 (below). Place the soil block a plastic sheet, then start hand sorting.

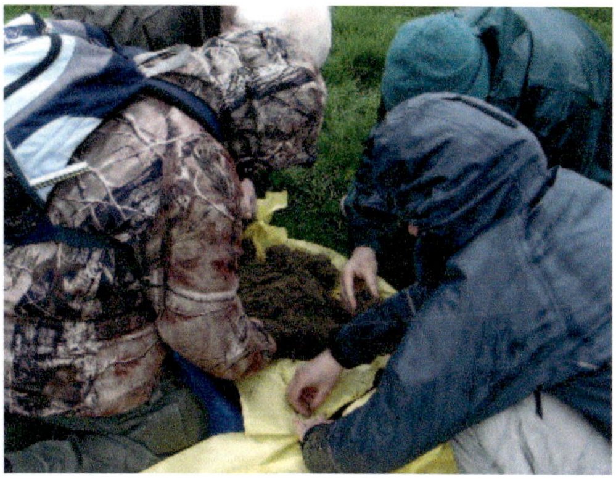

Figure 25. Hand sorting: break pieces of soil off from the block and finely sort through using fingers.

Figure 26. Microhabitats: searching in other earthworm niches including under leaf litter and stones.

Place the block on a tray or plastic sheet, such as a bin liner, and then start hand sorting. A good technique is first to break pieces off the block of soil, then very finely sort through using your fingers. Some earthworm adults are only a centimetre or two long. In diapause (a type of hibernation for earthworms when conditions are too dry) the worms can be curled up into tight knots. They can also be located in the roots of the grass.

Vermifuge

The best way to gather anecic worms is to use a vermifuge: an irritant poured on the ground. This technique is also useful in places where digging is not allowed, such as golf courses. Formalin is the most effective vermifuge, but since it is hazardous for you and would kill the worms it is not recommended (see Safety note on page 20). A solution of mustard powder dissolved in water is the best alternative, as long as the soil is not too waterlogged or extremely dry. Use 4 large heaped tablespoons in 2 litres of water for a 25 x 25 cm square. It can take up to ten minutes for the last of the worms to emerge. They may not all emerge in the square you poured the mustard solution into. Check that worms have emerged fully before picking them up. Do not pour the solution over too large an area as it needs to percolate right down into the soil. When juveniles have been counted wash them in clean water and release them back into the soil, also repeat for any adults not required for further study.

Microhabitats

Not all worms are found in the soil, so it is worth searching other earthworm niches. Look under leaf litter, stones, and any objects lying on the ground. Examine rotting logs, looking underneath and inside, especially under the bark and in any notches or holes in tree trunks.

Recording data and labelling

When collecting worms in the field, always record and label as you go along. The more information that is recorded about a specimen, the more valuable it is.

Useful equipment to take in the field

- Sampling tubes for collecting earthworms
- Paper labels to be placed with the worm inside the tube (preferably conservation grade such as Resistall; however good quality paper is fine)
- Pencil (although pencil can fade over time, it is fine in the field)
- Scissors
- GPS (or a compass and local map)
- Notebook (there are special waterproof notepads which are very useful) or data recording sheets which prompt you to fill in all details

Figure 27. Useful equipment to take in the field (Map data © OpenStreetMap contributors).

	A	B	C	D	F	F	G	H	I	J	K
1	Species	Site Name	Grid Reference	ViceCounty	Date (dd/mm/yyyy)	Recorder	Determiner	Number	Habitat	Method	Comment
2	Allolobophora chlorotica	Newcastle Park	XL3455678	South Wiltshire	27/02/2018	John Doe	John Doe	1	041 park	401 Soil pit sampling	Next to footpath
3	Aporrectodea longa	Newcastle Park	XL3455678	South Wiltshire	27/02/2018	John Doe	John Doe	2	041 park	402 Soil pit sampling	Next to footpath
4	Dendrobaena octaedra	Newcastle Park	XL3455678	South Wiltshire	27/02/2018	John Doe	John Doe	3	041 park	421 Under deadwood bark	Small log
5	Lumbricus rubellus	Newcastle Park	XL3455678	South Wiltshire	27/02/2018	John Doe	John Doe	4	041 park	455 Dung search	Under deer dung
6	Lumbricus terrestris	Newcastle Park	XL3455678	South Wiltshire	27/02/2018	John Doe	John Doe	5	041 park	401 Soil pit sampling	Next to footpath
7											
8											
9											
10											

Figure 28. Earthworm Records Submission Sheet used by the National Earthworm Recording Scheme (www.earthwormsoc.org.uk).

As soon as you get to your first site, write in the notebook or on your data sheet as much information as you can. Location must be recorded as precisely as possible (Ordnance Survey 6-figure grid references are used by the National Earthworm Recording Scheme). Habitat information is also important, so record for example whether you are in a woodland or grassland. Note which sampling method was used, such as found by searching under deadwood. Record the date.

Use a separate sampling tube for each individual microhabitat sampled (e.g. from a log, the soil, leaf litter, etc.), with its own label put in the tube.

Preservation techniques

If the worms have been brought to the lab live and are needed for a long-term reference collection, go straight to the 'in the lab' instructions on page 20. Otherwise follow one of the 'in the field' methods below.

In the field

Method 1: the quick way

For UK and Irish species, long-term storage for dissection is not required for identification, so this method is suitable if the worms are not needed for a reference collection. It is also useful if you are collecting worms primarily to gather molecular data (DNA work). You will need:

- Good quality tubes which won't crack
- Alcohol 70-80% (96% if needed for molecular work)

The quickest way to preserve worm samples is to place them straight into 70-80% alcohol as soon as they are collected. Do not fill the tube more than a quarter full with worms. Worms have a large quantity of fluid in them, so when they die they release those fluids and dilute the alcohol. Change the alcohol in the tube after 5 minutes and then again after 15 minutes. If worms are needed for DNA studies they should be placed in 100% alcohol (N.B. most absolute alcohol is actually about 96% rather than 100%, but this is fine). If not needed for molecular studies, it is best to use 70-80% alcohol to prevent the worms from becoming too brittle.

This method will keep your worms safe. However, they may end up bent in an awkward position which can make identification more difficult later. They will also have more soil particles attached to them, are at increased risk of becoming soft – as fluids first released from the worm in the process will be released straight into the tube and they will not be as firm and easy to handle as those preserved in the method below.

Method 2: best practice

You will need:

- Good quality tubes which won't crack
- Alcohol 70-80% (96% if needed for molecular work)
- Tupperware tub (leak proof)
- Water (use distilled water to dilute alcohol as tap water can make the solution cloudy)

Step 1: **Relaxation and anaesthetising**. Place the worms straight into a tub of 20% alcohol. This can be made up beforehand and carried with your kit as long as it is leak-proof. This will help 'clean the worms' as well as relaxing and anaesthetising them. Leave them in this for approximately 2 minutes – but no longer as the worms must not die in this completely relaxed state. When the worms stop moving take them out.

Step 2: **Straightening and fixing.** Worms can be straightened while they are fixing. Use 70-100% alcohol. 80% alcohol is fine unless DNA work might be required; in which case use 96-100% alcohol at all times.

Although the best straightening tool is a 'Pop' worm tray (see Fig. 30 and page 21 for instructions on how to make this) it is generally not practical for use in the field unless it is made very light. Instead worms can be straightened using a Tupperware tub lid – the tub used to hold your anaesthetising solution is ideal (Fig. 29). While the worms are still floppy from the anaesthetising, lie them between two straight objects (or lay them straight in the upturned lid of your Tupperware box, along the edge of the rim), then add the fixative of high percentage alcohol. Formalin is not recommended due to the health risks associated.

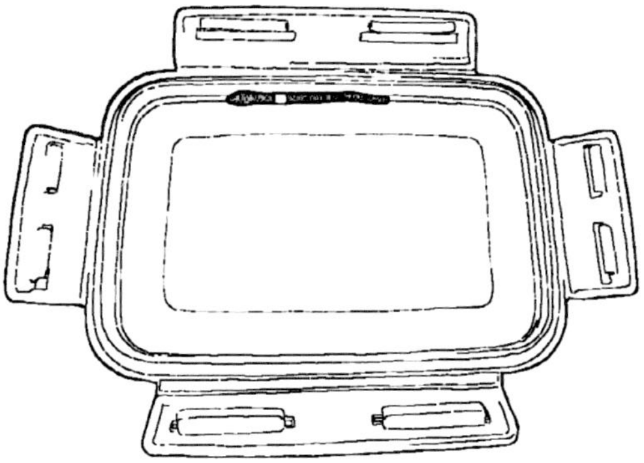

Figure 29. Using the lid of a Tupperware box to straighten earthworms.

Let the fixative pool a little around the worm. Ideally worms should be kept this way for a long period of time but since this is not possible in the field, place your worm in your collecting tube after 5-10 minutes. This should make identification easier than with the quick method above.

In the lab

Follow these procedures if the worms have been brought to the lab live. They include instructions on long-term preservation for a reference collection. You will need:

- Good quality tubes which won't crack
- Alcohol 70-80% (96% if needed for molecular work)
- Alcohol 20%
- Worm tray (optional)

Step 1: **Relaxation and anaesthetising.** Place the worms straight into a tub of 20% alcohol. This will help 'clean the worms' as well as relaxing and anaesthetising them. Leave them in this for approximately 5 minutes – but no longer as the worms must not die in this completely relaxed state. When the worms become more sluggish take them out.

Step 2: **Straightening and fixing.** Worms can be straightened while they are fixing. The best fixatives are:

1. 70-80% alcohol. 70% alcohol is fine unless DNA work might be required; in which case use 96-100% alcohol at all times
2. 4% formalin (* see Safety note below), if colour needs to be preserved but the worm will not be needed for molecular work

---

* **Safety note: Formalin.** Formalin is the best preservative but is hazardous to health. Follow the supplier's safety instructions carefully. Use a fume hood, and wear rubber or plastic gloves and safety glasses.

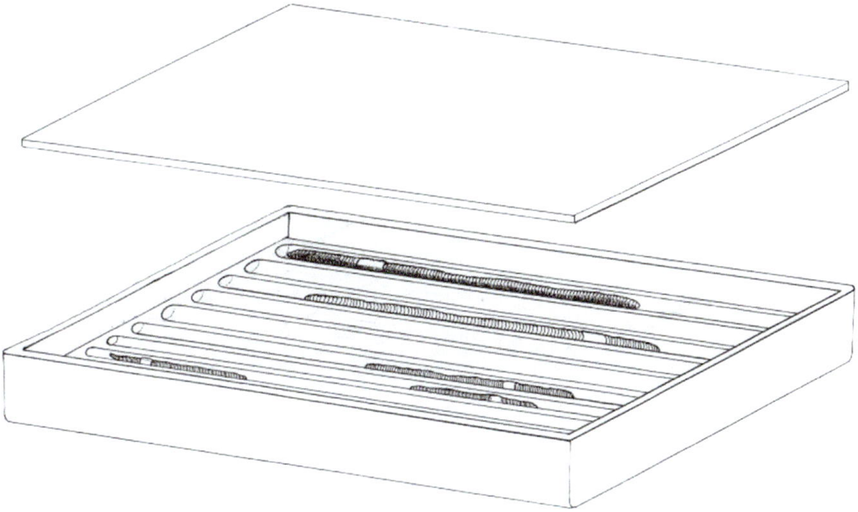

Figure 30. A 'Pop' worm tray in use.

The best straightening tool is a 'Pop' worm tray (designed by the Victor Pop senior in Romania). This is a block of wood or plastic with furrows chiselled into it, which can be made with a soldering iron on plastic or a file on wood. Worms can be fixed within the furrows of the tray by pouring the fixative over the tray and covering with a lid to prevent the worms from drying out.

Leave the worms in the fixative for as long as is possible, ideally at least 2 days. Then transfer them to the preservative. 80% alcohol is fine for long-term storage. If it is possible to have a few months to a year in 4% formalin first, this will help the colour of the worm to be preserved in the longer term and keep the worm firm, although this is not always possible due to the health risks associated with formalin.

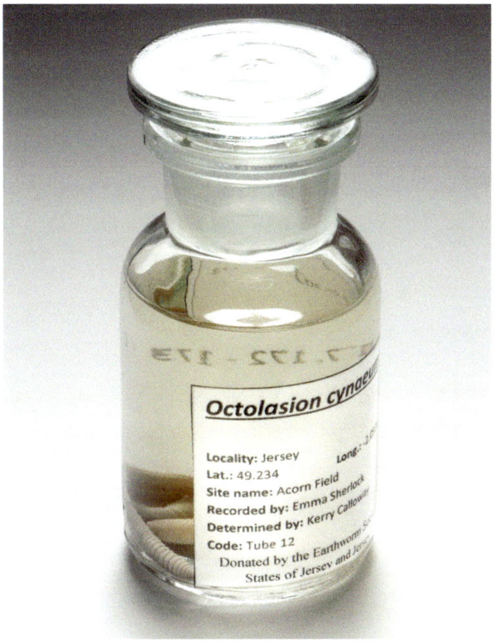

Figure 31. Long-term storage of an earthworm specimen.

Long-term storage

For long-term storage place the worms in either:

1. 80% alcohol

2. 4% formalin (* see Safety note on page 20)

as outlined on page 19.

The best long-term storage jars are made of glass and have ground glass stoppers. Suitable glass jars can be sourced from Dixon Science (www.dixonscience.co.uk). Grease the seal with vaseline or soft white paraffin.

Using the key to earthworms found in natural environments

Only adult worms can be accurately identified to species using this key.

To distinguish an adult worm from a juvenile look for a clitellum (or saddle). If no clitellum is present the specimen is a juvenile.

If there is a TP present but no clitellum the specimen is a sub-adult. It is sometimes possible to identify sub-adult worms from the shape and position of TP, male pore, setae and head by using the Quick comparison chart (pages 40-41) but this is only recommended for experienced users.

If the clitellum is not fully developed but some markings are visible that could be a TP (see explanation on page 26) the specimen is a juvenile earthworm which could be identifiable.

You will need to be familiar with where on an earthworm's body the features highlighted below are to be found. Refer to the Biology section (pages 9-13) where the external anatomy is highlighted if you are unsure.

Microscope work

A low powered microscope or stereomicroscope is needed for identifying earthworms. Ideally the microscope should be capable of low magnification (x10) as well as going up to x30 or x40 magnification for studying setae on small worms. Having the capability of just x10 magnification makes a big difference when studying larger worms. Make sure the microscope is set up correctly before starting.

Place the preserved worm in a plastic or glass clear Petri dish and always completely immerse the worm in water or alcohol. When water is being used, leave the water in a bottle overnight if possible so any bubbles have settled.

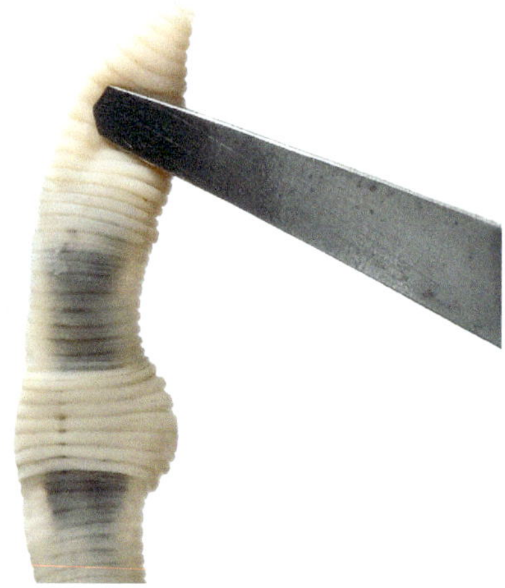

Figure 32. Use forceps to get a good view.

When looking at a particular character, especially setae, which can be difficult to locate, manipulate the specimen and light sources to find the best angle for viewing. Features which cannot be seen at first can become quite apparent after changing the angle of view or direction of the light. Some microscopes will also allow you to switch the background plate between white and black, which may help to improve contrast with the specimen.

Forceps and seekers are the best tools for manipulating or turning the worm (see Fig. 32). Do check both sides of the worm when viewing a structure such as the TP – sometimes one side is clearer than the other.

Main features explained

Head and tail; dorsal and ventral views

To distinguish between the head and tail, look for the position of the clitellum (saddle). The end closer to the clitellum (and the male pore) is the head. The other end is the tail.

The dorsal surface is the upperside of the earthworm, while the ventral surface is the underside. On the dorsal surface the clitellum is raised higher above the level of the surrounding segments than on the ventral surface, where the clitellum is almost flush with the rest of the body. In newly preserved specimens that retain some pigment, the dorsal side may often be darker than the ventral side.

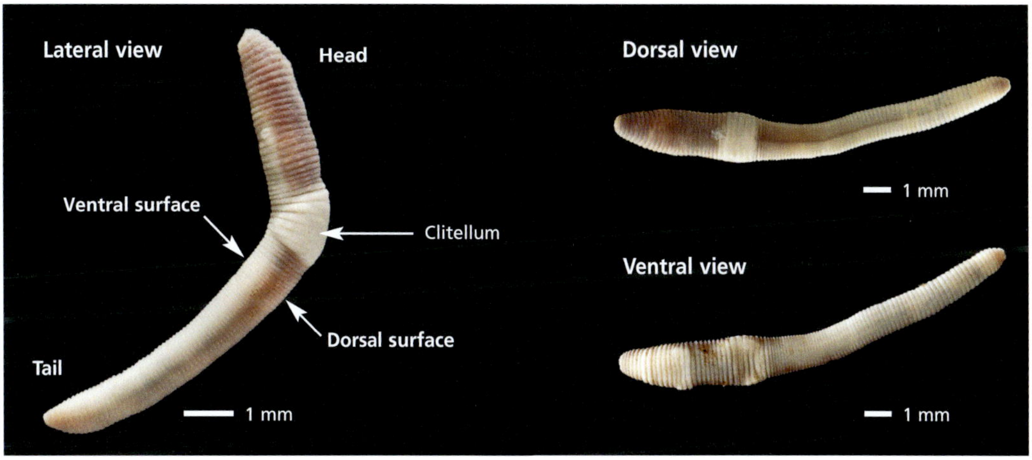

Figure 33. Distinguishing dorsal and ventral views.

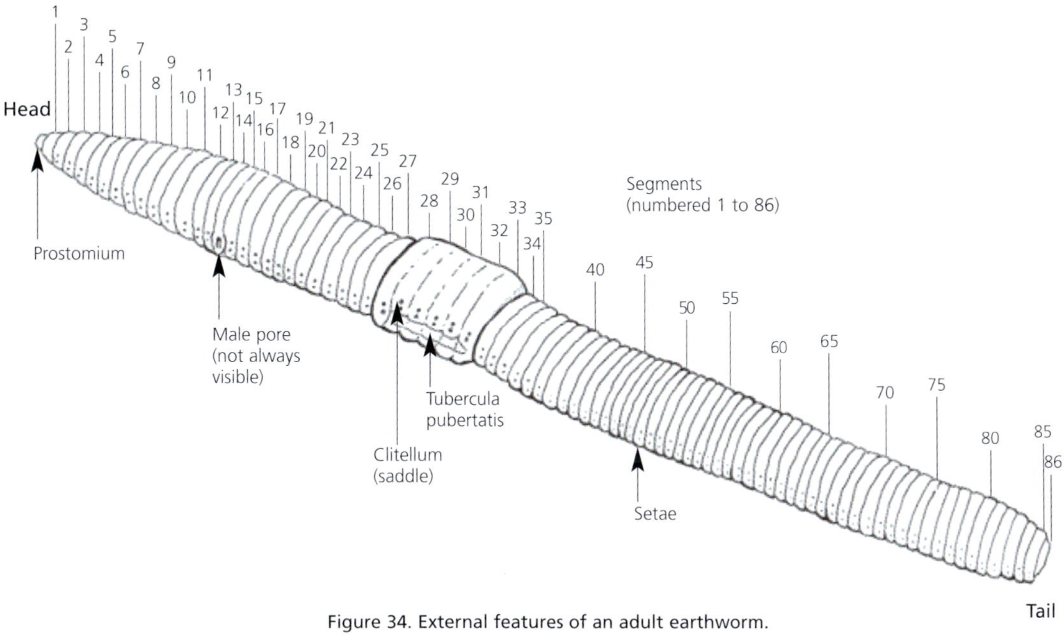

Figure 34. External features of an adult earthworm.

Counting segments

You will often have to do this. View the dorsal surface to get started. Count from the head end (see Fig. 35). The first segment to count is the first full ring. Do not count the prostomium.

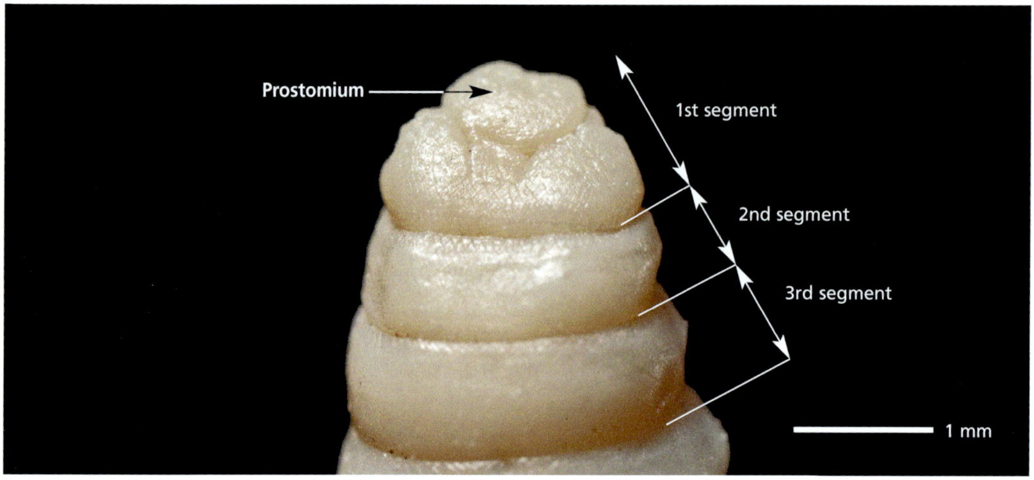

Figure 35. Counting the first few segments from the head.

Some segments can have confusing furrows through them. If unsure, look for the setae ridges to count segments.

Be careful when counting segments on the clitellum – the segments often become wider.

Figure 36. Segments can become wider on the clitellum.

Tanylobic or epilobic head

View the head dorsally to see this character. A tanylobic head is where the head-line of the prostomium meets the 2nd segment. An epilobic head is where the head-line of the prostomium does not meet the 2nd segment but stops short.

Sometimes creases in the skin can resemble the furrows associated with a tanylobic head. Figure 39 may look like a tanylobic head but the lines do not reach the 2nd segment.

If unsure key out in both sides of the key and check species accounts pages carefully.

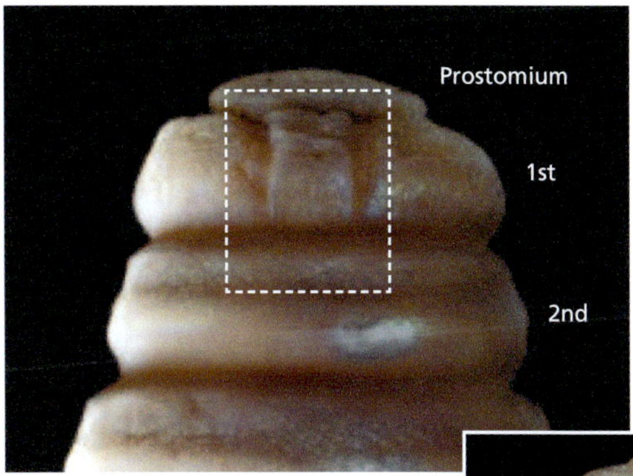

Figure 37. Tanbylobic head: the head-line of the prostomium meets the 2nd segment.

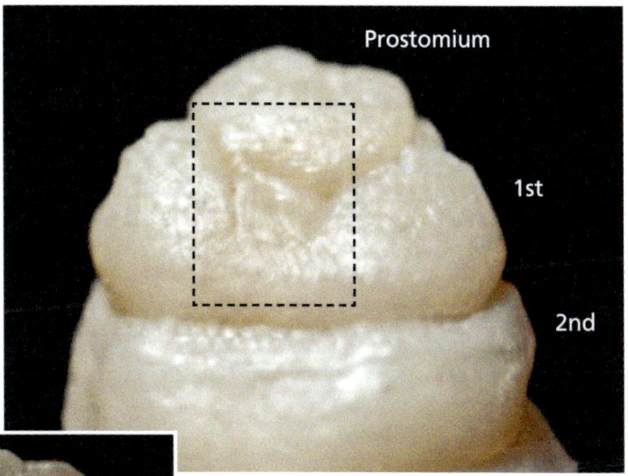

Figure 38. Epilobic head: the head-line of the prostomium does not meet the 2nd segment.

Figure 39. Example of an epilobic head which might be mistaken for a tanylobic head.

Figure 40. Tanbylobic head: side view.

Figure 41. Epilobic head: side view.

Clitellum

The presence or absence of a clitellum (or saddle) indicates whether it is a juvenile or adult worm.

The positioning of the saddle is very important when identifying worms.

Figure 42. Adult worm. A clitellum is present.

Figure 43. Juvenile worm. No clitellum is present.

Tubercula pubertatis (TP)

These are the markings, lumps, lines, swellings or 'sucker like discs' that are to be found on the lateral undersides of the clitellum. They can be very diverse.

Make sure that you check both sides, as they can be clearer on one side than the other. Sometimes it helps to locate the TP with the naked eye before examining it under the microscope.

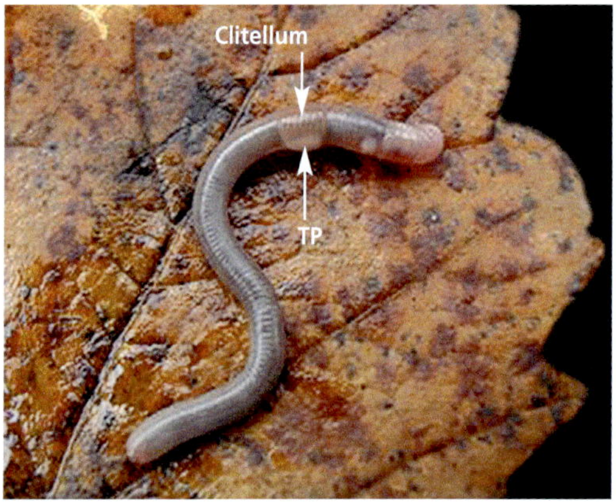

Figure 44. Position of tubercula pubertatis (TP).

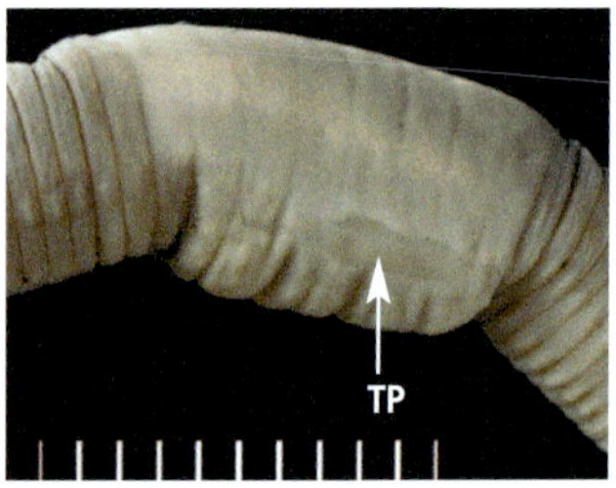

Figure 45. *Aporrectodea longa*.

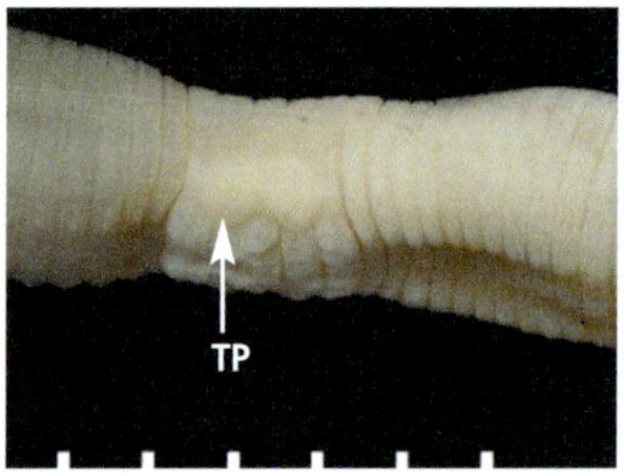

Figure 46. *Aporrectodea calignosa*.

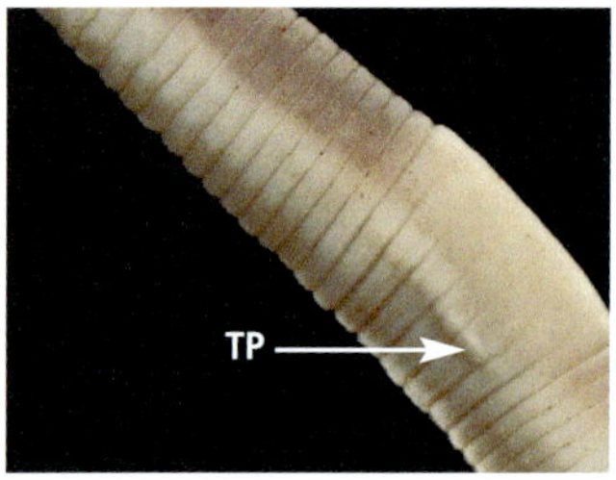

Figure 47. *Bimastos rubidus*.

Male pore

In the Lumbricidae family this pore is found on segment 13, 14 or 15 (Figs 48-50). It can be very large encompassing the surrounding segments. It can even look a bit more like a slit than a pair of small lips (Fig. 49). Sometimes it is very hard to see (Fig. 50).

Figure 48. *Allolobophora chlorotica* has a male pore like a pair of lips.

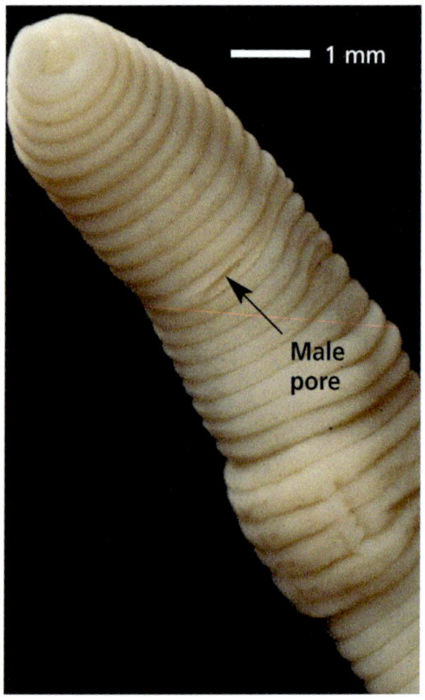

Figure 49. *Eiseniella tetraedra* has a male pore shaped like a slit.

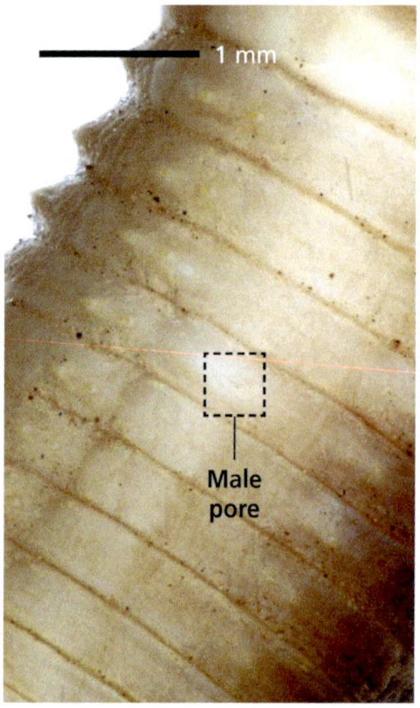

Figure 50. *Lumbricus rubellus* has a male pore which is hard to see.

Setae

Setae are small bristles or hairs. Earthworms all have small setae on each segment, and in the Lumbricidae family there are 4 pairs of setae per segment. Whether the setae are widely paired or closely paired after the clitellum can be useful diagnostically.

This can be the hardest character to see. Often setae can only be seen as small black dots or lines down the worm's body. It is crucial that the worm is clean, as specks of dirt can mask the setae. The setae should be viewed after the clitellum. They are often easisest to discern right at the tail end. Adjusting the microscope lighting and rotating the worm can help.

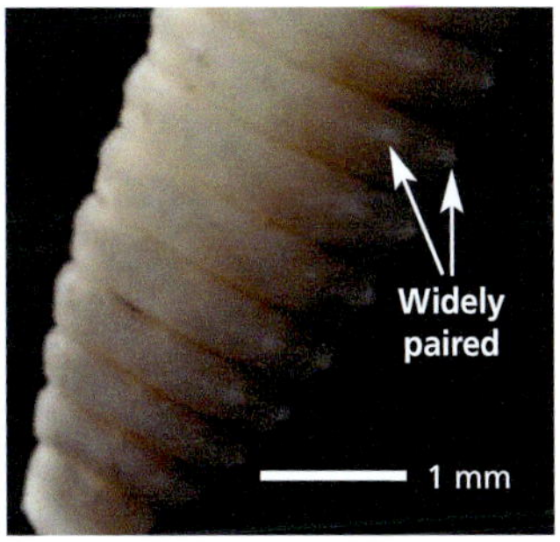

Figure 51. *Dendrobaena octaedra*.

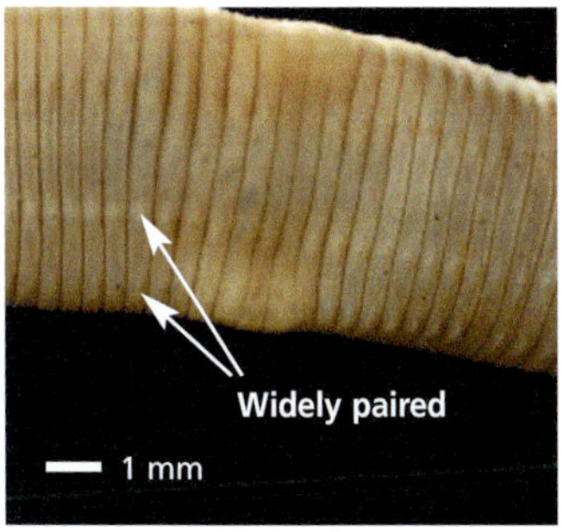

Figure 52. *Dendrobaena veneta*.

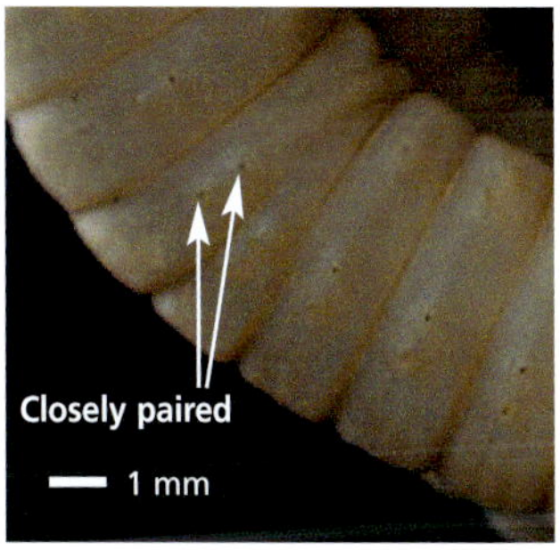

Figure 53. *Lumbricus terrestris*.

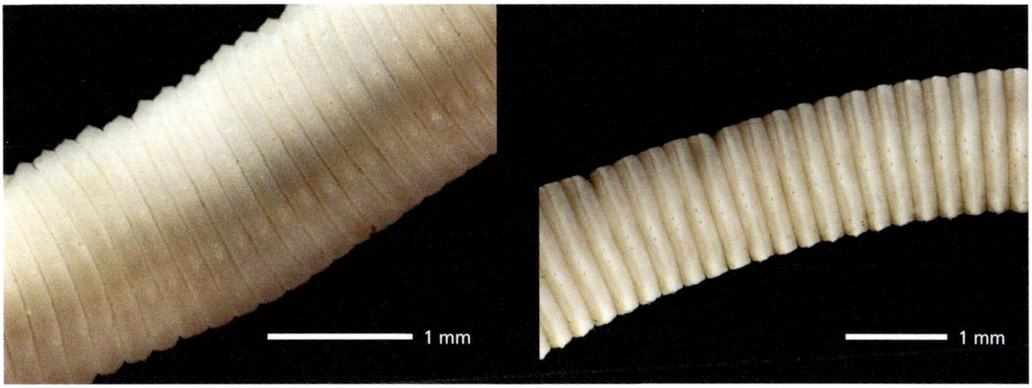

Figure 54. Setae.

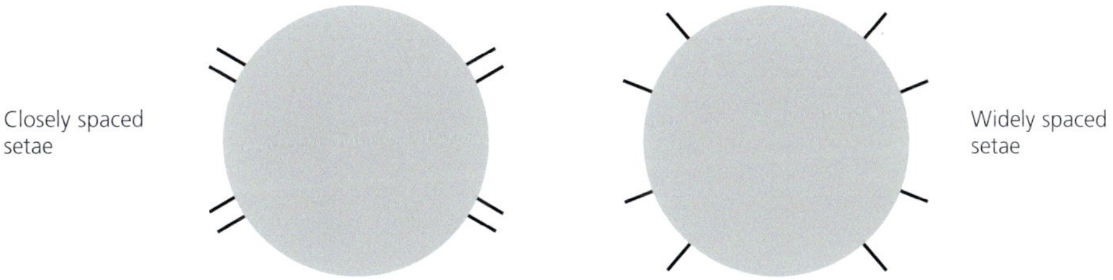

Figure 55. Cross sections to show setae spacing.

Key to earthworms found in natural environments

If identifying earthworms for the first time, consult the section 'Using the key to earthworms found in natural environments' (p. 22) before getting started.

1. Is the head tanylobic or epilobic?

View the earthworm dorsally for this character – see page 23.

Tanylobic .... Go to Question 2

Head-line of prostomium meets the 2nd segment.

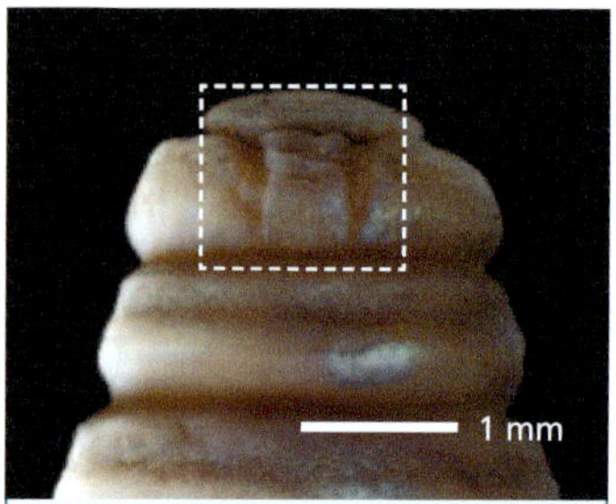

Check: When colour is visible, worms with tanylobic heads are red. If in life your worm was not red in colour you have gone wrong – check the head again.

– Epilobic .... Go to Question 7

Head-line of prostomium does not meet the 2nd segment but stops short.

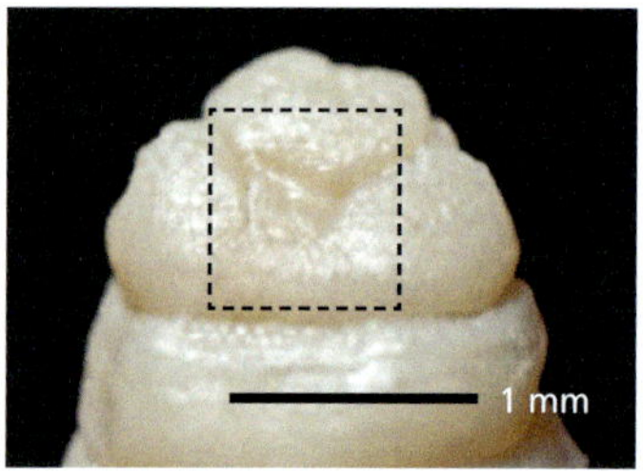

2. Which segment does the clitellum start on?

It is best to view the earthworm dorsally for this character.

– Segment 29 or before .... Go to Question 3

– Segment 30 or after .... Go to Question 6

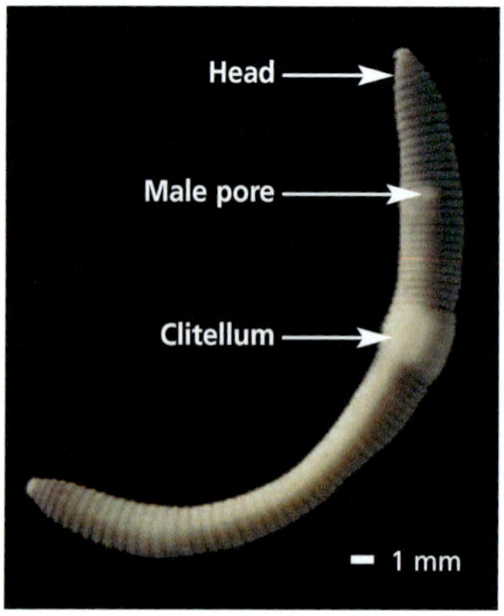

3. Is there a TP?

TP can be seen on lateral side of clitellum. It is best to view the earthworm ventrally for this character.

- Yes, the TP is present .... Go to Question 4
- No, the TP is absent .... *Bimastos eiseni* (p. 51)

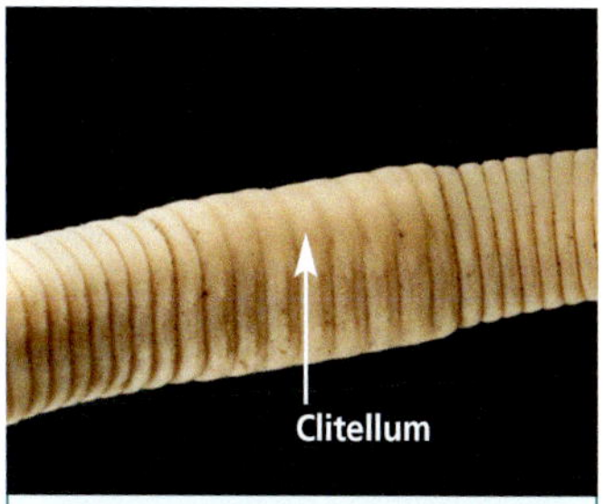

Check: *B. eiseni* clitellum starts on segment 24 or 25.

4. Are the setae closely paired?

Setae are often easier to see towards the tail. Always look after the clitellum, never at the head end. This character is best seen on a black background if present on your microscope.

- Closely paired .... Go to Question 5
- Widely paired .... Go to Question 28

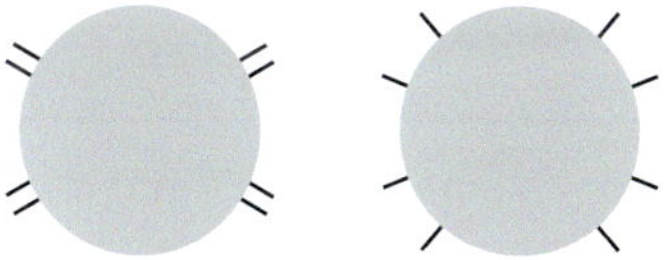

5. Which segments does the TP fall on?

- Segments 27/28 to 31 .... *Lumbricus rubellus* (p. 65)

  Occasionally the TP of *L. rubellus* can have a shadow underneath with a shape that resembles a pair of spectacles. It is the raised area above this that constitutes the TP in this case.

- 29 to 32 .... *Lumbricus castaneus* (p. 62)

6. Which segments does the TP fall on? (Count very carefully)

Read all 3 options

– Segments 32/33 to 36/37 .... *Lumbricus terrestris* (p. 66)

– Segments 34 to 36 .... *Lumbricus friendi* (p. 64)

– Segments 35 to 38 .... *Lumbricus festivus* (p. 63)

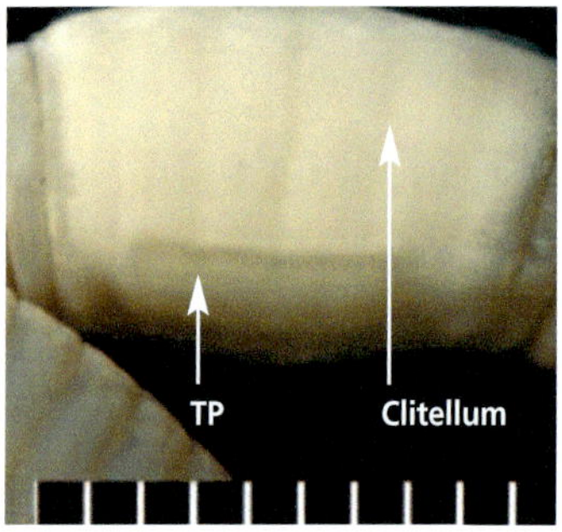

Lumbricus terrestris.

7. Which segment does the clitellum start on?

– Segment 20 or before .... Go to Question 8

– Segment 21 of after .... Go to Question 9

8. Which segment does the clitellum start on?

Read all 3 options

– Segment 13 .... *Microscolex phosphoreus* (p. 67)

– Segments 14 or 15 .... *Sparganophilus tamesis* (p. 72)

Check: *Sparganophilus tamesis* is only found in very waterlogged soils and at the bottom of rivers.

– Segment 19 .... *Kenleenus armadas* (p. 61)

Check: *Kenleenus armadas* has only been found in Ireland.

If none of these re-examine the specimen.

9. Is the male pore on segment 13 (or rarely 14)?

– Yes .... *Eiseniella tetraedra* (p. 59)

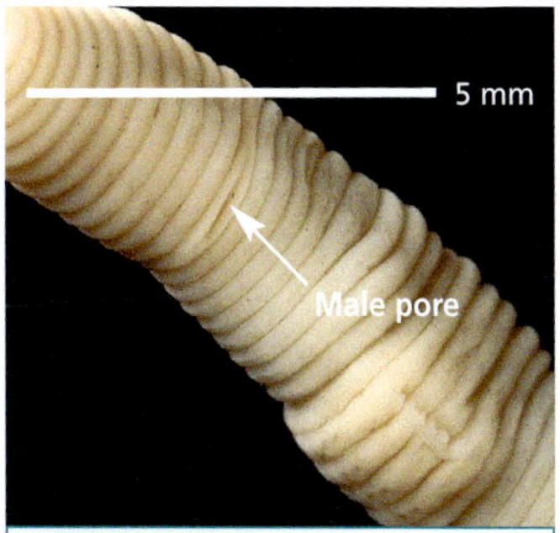

Check: The male pore of *E. tetraedra* can be small and look like an extra furrow in the segment.

– No .... Go to Question 10

> **Check:** If you cannot see a male pore AND the earthworm is large AND red, re-check whether the head is epilobic
>
> If the head still looks epilobic and the worm is large and red with closely paired setae and a TP running from 28 to 31, and you did not collect this worm in a compost bin, it is likely you have a malformed *Lumbricus rubellus* (p. 65).

10. Which segment does the clitellum start on?

– Segment 30 or before .... Go to Question 11

– Segment 31 or after .... Go to Question 31

11. Is there a TP present?

TP can be seen on lateral side of clitellum. It is best to view the earthworm ventrally for this character.

– Yes, the TP is present .... Go to Question 13

– No, the TP is absent .... Go to Question 12

12. Are the setae closely paired?

Setae are often easier to see towards the tail. Always look after the clitellum, never at the head end. This character is best seen on a black background if present on your microscope.

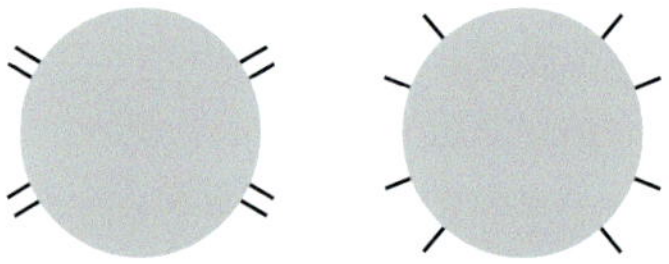

– Closely paired .... *Murchieona muldali* (p. 68)

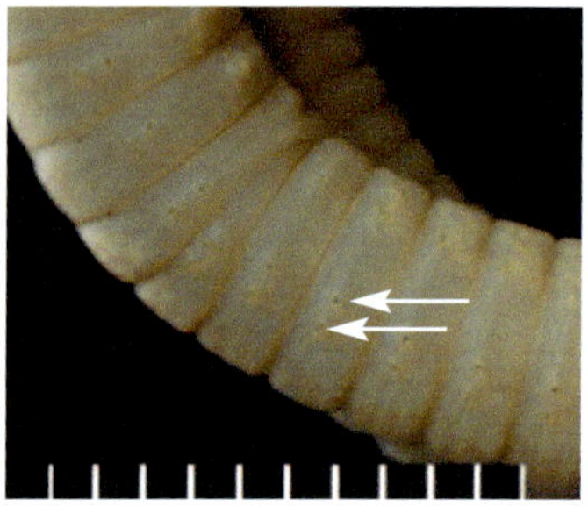

Check: *M. muldali* is a very small grey-pink worm. The clitellum is on segments 27 to 33. It is usually just 1.5-4 cm long. If your worm does not match this description then the TP may just be hard to see. Please check the worm again.

– Widely paired .... *Bimastos rubidus* (p. 52)

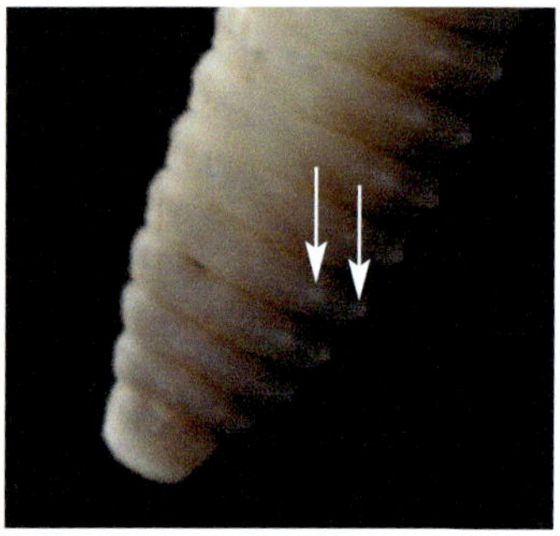

Check: A rarer form of *Bimastos rubidus*, in which the TP is absent or very hard to see, keys out here. It is a red worm with its clitellum on segments 25-27 to 31-32(33), and setae widely spaced. See page 52 for more information.

13. Does the TP start on segment 31?

– Yes, the TP starts on segment 31 .... Go to Question 14

– No, the TP starts elsewhere .... Go to Question 19

14. Does the TP end on segment 33?

– Yes, the TP ends on segment 33 .... Go to Question 15

– No, the TP ends elsewhere .... Go to Question 16

15. Which of these descriptions best matches the TP?

- A thin band on segments 31 to 33 (can sometimes be an indistinct raised area) .... *Dendrobaena octaedra* (p. 55)

  In life a small red worm.

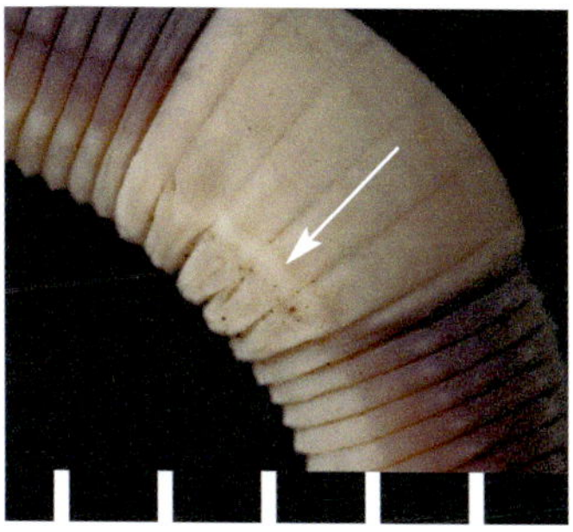

D. octaedra.

- 2 'humps' over the area of segments 31 to 33 .... Go to Question 18

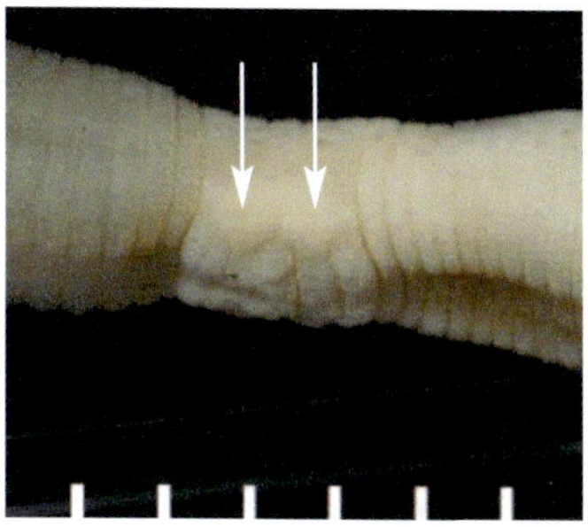

2 'humps' on 31 to 33.

16. Is the TP shaped like 3 'sucker like discs' (or 3 distinct lumps) on segments 31, 33 and 35?

- Yes, the TP matches that description .... *Allolobophora chlorotica* (p. 43)

  In life a green or pink worm.

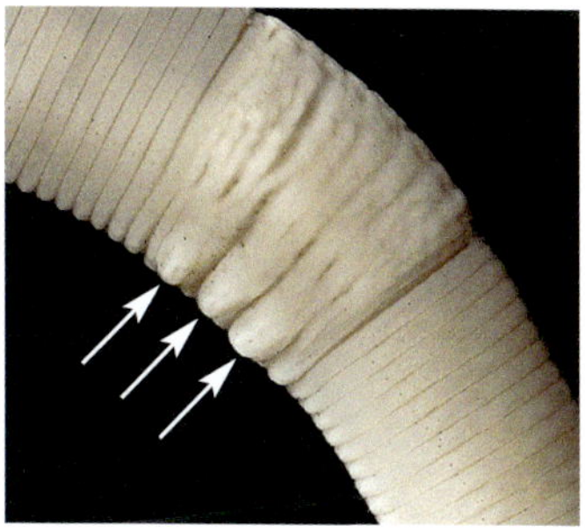

A. chlorotica.

- Not, the TP is not like that .... Go to Question 17

17. Are the setae closely paired?

Setae are often easier to see towards the tail. Always look after the clitellum, never at the head end. This character is best seen on a black background if present on your microscope.

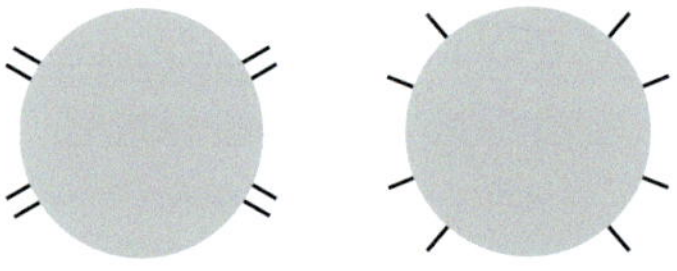

– Closely paired .... *Aporrectodea longa* (p. 48)

Check: *A. longa* is a dark headed worm.

– Widely paired .... *Octolasion lacteum* (p. 70)

Check: *O. lacteum* is a paler pink-grey worm.

18. Read both check boxes

You are likely to have found *Aporrectodea caliginosa.*

Check: *Aporrectodea caliginosa* is one of the commonest earthworms in the UK and Ireland (p. 44). This is an endogeic worm. It is grey or pink in colour in life. It is very quick to lose its colouration when preserved in alcohol.

A. caliginosa.

But there is a small chance that you have found *Aporrectodea nocturna.*

Check: *Aporrectodea nocturna* (p. 49) is an anecic worm. It has a deep red head in life. Its head remains deep red when preserved in alcohol, although the rest of the body becomes pale.

A. nocturna.

19. Are the setae closely paired?

Setae are often easier to see towards the tail. Always look after the clitellum, never at the head end. This character is best seen on a black background if present on your microscope.

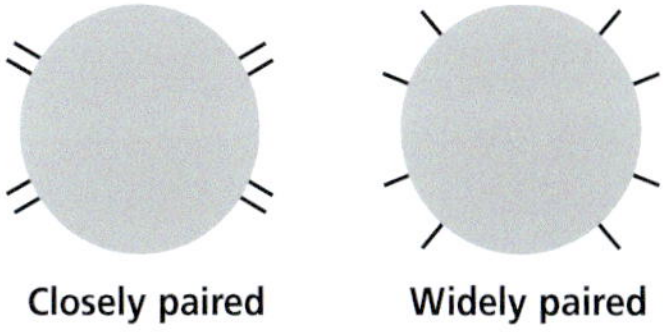

– Closely paired .... Go to Question 20

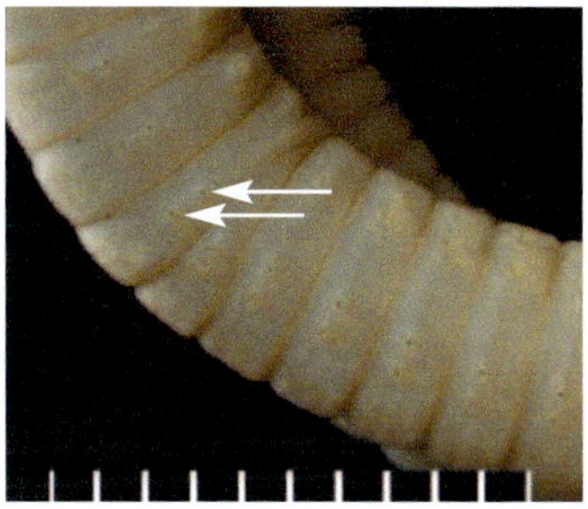

– Widely paired .... Go to Question 26

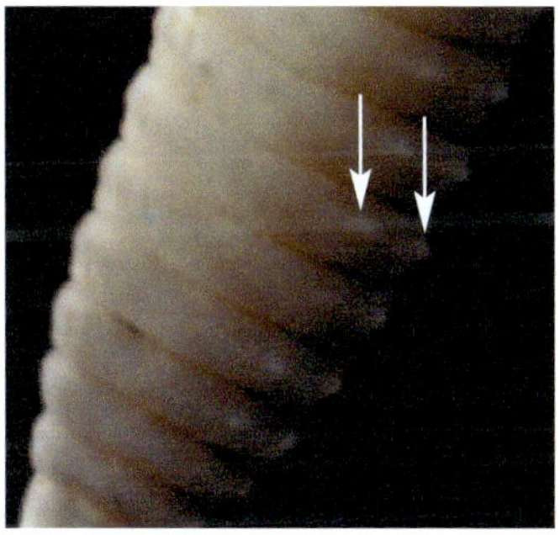

20. Which segment does the TP start on?

– Segment 31 or after .... Go to Question 21

– Segment 29 or before .... Go to Question 22

21. Which segments does the TP fall on?

– Segments 32 to 34 .... *Aporrectodea longa* (p. 48)

A. longa, the TP is a broad band.

– Segments 33 to 34 .... *Aporrectodea limicola* (p. 47)

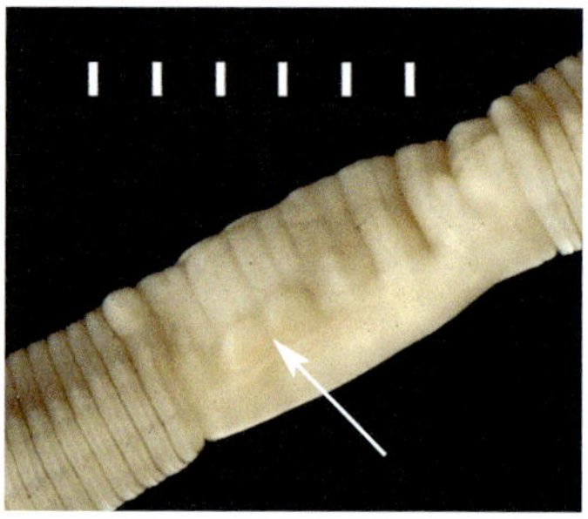

A. limicola, the TP is 2 humps.

22. Which segment does the TP start on?

– Segment 27 or 28 .... Go to Question 23

– Any segment other than 27 or 28 .... Go to Question 24

Check: Check carefully here that the TP starts on these segments and not 29.

23. Does the TP have 2 humps or suckers (predominantly on segments 28 and 30)?

– Yes, the TP matches that description .... *Aporrectodea cupulifera* (p. 45)

Check: *A. cupilifera* has so far only been recorded in Ireland and not in Great Britain.

– No, the TP is different .... *Eisenia fetida* (p. 58)

This is a composting worm rarely found outside of the compost bin, dead wood or areas rich in organic matter.

24. Which segment does the clitellum start on?

- Segments 21, 22 or 23 .... Go to Question 25
- Segments 24, 25 or 26 .... *Aporrectodea rosea* (p. 50)

  A. rosea has a TP on segments 29-31 (occasionally just 29-30).

Check: If the TP is not visible and the clitellum falls on segments 26/27-33/34, you most probably have *Murchieona muldali* (p. 68).

25. Which segment does the TP start on?

- Segment 24 or before .... *Eiseniella tetraedra* (p. 59)

  (a rarer form with the male pore on 15 not 13)
- Segment 25 or after .... *Helodrilus oculatus* (p. 60)

Check: Both *Eiseniella tetraedra* and *Helodrilus oculatus* are usually found in wet soils.

Check: If the clitellum starts on 23 and the worm has a flared TP 29-30 (31) then this might be a rare form of *Apporectodea rosea.* Check species page (p. 50).

26. Which segment does the clitellum start on?

- Segments 25, 26 or 27 .... Go to Question 27
- Segments 28, 29 or 30 .... Go to Question 29

27. Does the TP end on segment 30?

- Yes, the TP ends on segment 30 .... *Bimastos rubidus* (p. 52)
- No, the TP ends elsewhere .... Go to Question 28

28. Use the table to help you identify the species

Dendrobaena veneta and *Dendrobaena hortensis* can often be distinguished by size, but can be difficult to tell apart if they fall within the overlapping size range. There are also differences in colour (when living), adult body length and habitat.

	Dendrobaena veneta	*Dendrobaena hortensis*
Colour	Striped red-violet colouration; paler ventrally	Reddish colouration
Adult length range	30-155 mm	15-50 mm
Adult diameter range	4-8 mm	1.5-5 mm
Habitat	Mainly compost heaps but also under leaf litter and places with high organic matter	Mainly very organic-rich soils and forest leaf litter; can also be found in compost

29. Does the TP end on segment 32?

- Yes .... *Dendrobaena attemsi* (p. 53)
- No .... Go to Question 30

Check: *D. attemsi* is a small worm usually found in acid woodlands.

30. Which segments does the clitellum span?

- Segments 29 to 34 .... *Octolasion cyaneum* (p. 69)
- Segments 30 to 35 .... *Octolasion lacteum* (p. 70)

31. Which segments does the TP start on:

- Segments 33 or 34 .... *Satchellius mammalis* (p. 71)
- TP absent or not starting on 34 .... Go to Question 32

Check: *Satchellius mammalis* is a small red worm with a semi-circular TP.

32. Are the setae closely paired?

Setae are often easier to see towards the tail. Always look after the clitellum, never at the head end. This character is best seen on a black background if present on your microscope.

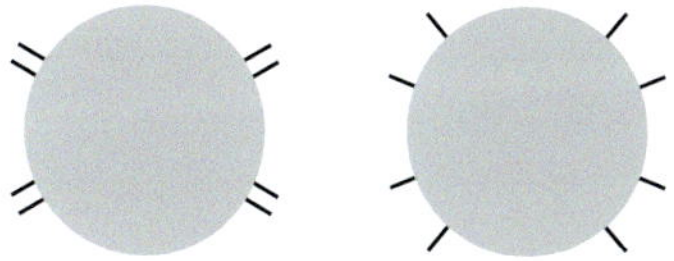

Closely paired **Widely paired**

– Closely paired .... *Aporrectodea icterica* (p. 46)

– Widely paired .... *Dendrobaena pygmaea* (p. 56)

> **Check:** *A. icterica* is a greyish worm with a clitellum on segments 33-34 to 42-44. *D. pygmaea* is a very small reddish worm with a clitellum on segments 32-33 to 37-38.

Quick comparison chart

Species	Head*	Setae	Colour (can vary)	Clitellum start	Clitellum end	TP	TP shape
Allolobophora chlorotica	epi	close	pink or green	28-29	37	31, 33, 35	3 mounds or discs 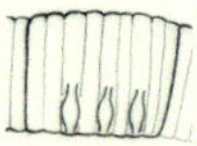
Aporrectodea caliginosa	epi	close	usually pink-grey	usually 29 (or as early as 25)	34-35	31-33	2 humps 
Aporrectodea cupulifera	epi	close	pink-grey	25-26	32	28, 30 (or part 27-31)	2 humps or discs 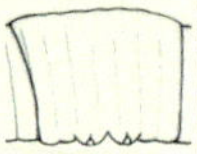
Aporrectodea icterica	epi	close	usually grey	33-34	42-43 (rarely 44)	35-36 to 42-43	long thin band 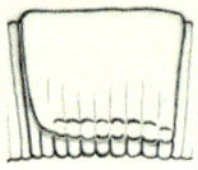
Aporrectodea limicola	epi	close	usually grey	28-29	35-36	33-34	2 humps 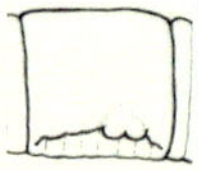
Aporrectodea longa	epi	close	very dark head	27-28	35-36	usually 32-34 (or 31)	band 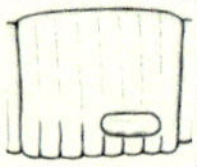
Aporrectodea nocturna	epi	close	red-brown anteriorly	Usually 29 (or as early as 25)	34-35	31-33	2 humps 
Aporrectodea rosea	epi	close	grey-pink	23-26	32-33	29-31 or only 29-30	raised band 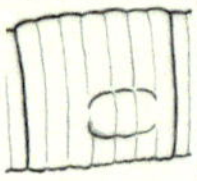
Bimastos eiseni	tany	close	dull red	27 usually (can be 25 or 26)	32-33	not present	
Bimastos rubidus	epi	wide	red	25-27	32-33, (rarely 31)	28-29 to 30 (or absent)	swelling 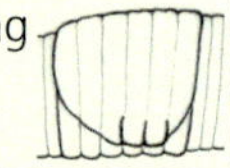
Dendrobaena attemsi	epi	wide	rosy to red	28-29	34	30-32	raised band 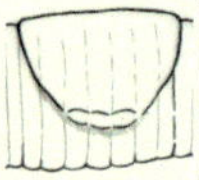
Dendrobaena hortensis	epi/ tany	wide	red	26-27	33	30-31 (or on 29 or 32)	swelling 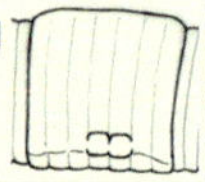
Dendrobaena octaedra	epi	wide	red	27-29	33-34	31-33	usually a thin band 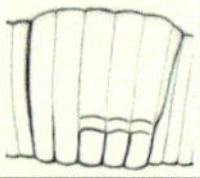
Dendrobaena pygmaea	epi	wide	variable	32-33	37-38	usually absent	ridge (if present) 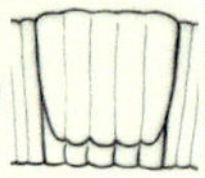

* epi = epilobic head
tany = tanylobic head

Species	Head*	Setae	Colour (can vary)	Clitellum		TP	TP shape
				start	end		
Dendrobaena veneta	epi/ tany	wide	red	26-27	32-33	30-31 (or on 29 or 32)	swelling 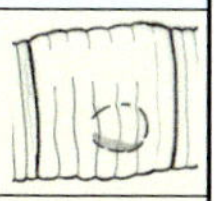
Eisenia fetida	epi	close	red	24-27	usually 32 (can be 33 or 34)	27-28 to 30-31 (can be 32)	thin bands 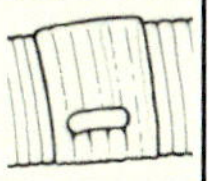
Eiseniella tetraedra	epi	close	pale	22-23	26-27	22-23 to 26-27	thin band 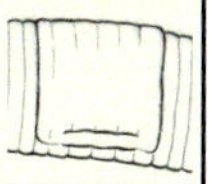
Helodrilus oculatus	epi	close	pale	21-22	32	29 to 30	ridge 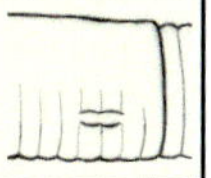
Keenleenus armadus	epi	wide	pale	half way across 19	28 (can be 29)	19-21 (or 22)	wavy ridge 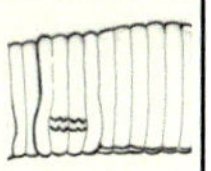
Lumbricus castaneus	tany	close	red	27-28	33-34	29-32	band 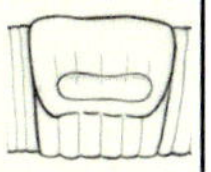
Lumbricus festivus	tany	close	red	33-34	40	35-38	band 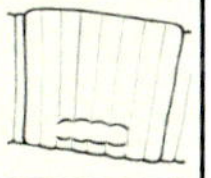
Lumbricus friendi	tany	close	red	32-33	37-38	34-36	thin ridge 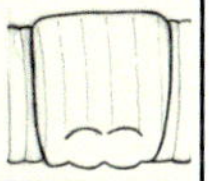
Lumbricus rubellus	tany	close	red	26-27	32	28-31 (can start 27)	thick band 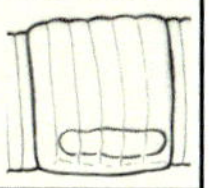
Lumbricus terrestris	tany	close	red	31 or 32	37	32-33 to 36-37	ridge 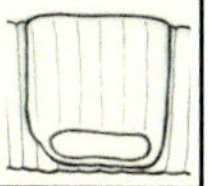
Microscolex phosphoreus	epi		usually light	13	17	not present	
Murchieona muldali	epi	close	pink-grey	26-27	33-34	unlikely to be present	
Octolasion cyaneum	epi	wide after saddle	blue-grey	29	34	29-30 to 33-34	band 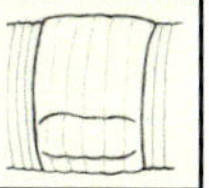
Octolasion lacteum	epi	wide after saddle	variable usually greyish	30	35	30-31 to 34-35	band 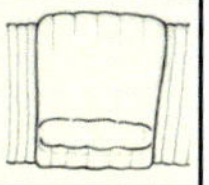
Satchellius mammalis	epi	wide	red	31	36	33 to 34	semi-circular band 
Sparganophilus tamesis	epi	close	pink	15 (rarely 14)	26	15-19 to 22-24	variable 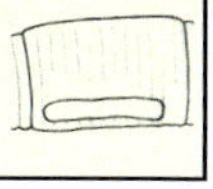

* epi = epilobic head
tany = tanylobic head

Species accounts for earthworms found in natural environments

31 species are included in the following pages. The photographs are taken from specimens housed at the Natural History Museum London or live photographs of worms collected on fieldwork, except where stated.

The preserved specimens loose colour so the natural colour is not maintained. The colour of the worm in life is noted and a live picture is attached where possible. A scale bar is included on all preserved pictures. Each white line represents 1 mm unless otherwise indicated.

Line drawings are the work of Louise Berridge.

The information included on the species pages was drawn together from a number of sources incuding the Earthworm Society of Britain, a Natural England report (Jones & Eggleton, 2014), the collections of the Natural History Museum, Sims & Gerard (1999), Csuzdi & Zicsi (2003) on the earthworms of Hungary, and Lowe & Butt (2008).

An indication is given as to how rare or common the earthworm is. Note though that still not enough is known about earthworm distributions around the UK and Ireland (Carpenter *et al.*, 2011). You can help change this by submitting records to the Earthworm Society of Britain!

Allolobophora chlorotica
(Savigny, 1826)

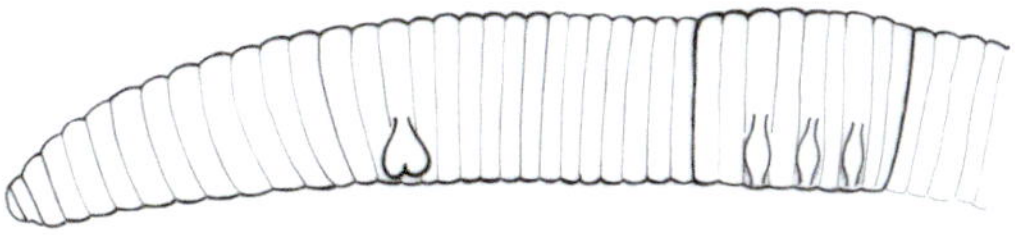

Very common

Ecological group: endogeic

Colour in life: two colour morphs exist: pink and green (A)

Habitat: Most habitats, particularly disturbed areas. Frequent in pastures, gardens and forests but prefers wet highly organic soils, neutral to base-rich grasslands and arable soils. Not usually found in the most acidic woodlands such as pine. Pink morph prefers drier conditions, mainly woodlands and gardens. Green morph prefers wetter conditions, dominant over pink in grasslands (Lowe & Butt, 2008). Biliverdin in the cells of the coelom produces green colouration

Cast: underground

External characters (B) (C)

Male pores: segment 15, encroaching onto 14 and 16

Clitellum: starts usually 29 (but rarely on 28) through to 37

TP: 3 mounds or lumps ('sucker like discs') on segments 31, 33 and 35 (C)

Setae: closely paired

Head: epilobic

Length: 30-80 mm. **Diameter:** 2-7 mm

Fecundity: usually one individual per cocoon. Incubation time 34-38 days at 20°C. Adult produces 10-27 cocoons a year (Lowe & Butt, 2005)

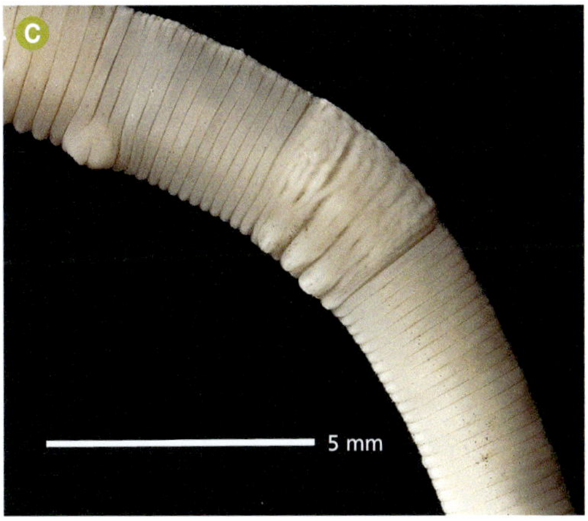

Internal characters

Seminal vesicles: 4 pairs segments 9-12

Spermathecae: 3 pairs 8/9 9/10 10/11

Last pair of hearts: in 11

Califerous glands: in 10-12 with lateral pouches

Aporrectodea caliginosa

(Savigny, 1826)

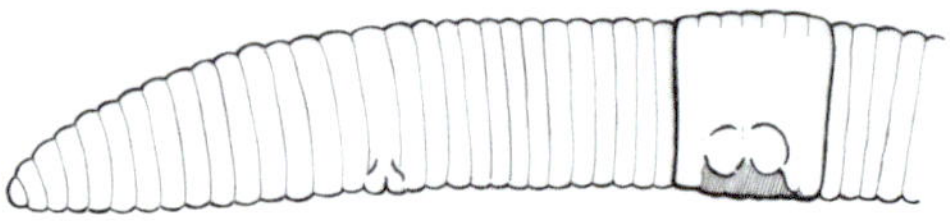

Very common

Ecological group: endogeic

Colour in life: pink/grey (A)

Habitat: Disturbance-tolerant species found in gardens, pastures, forests and arable fields. Very common in grassland. Not found in the most acidic habitats such as heaths and mires

Casts: Mainly underground but can also cast on the surface

External characters (B) (C) (D)

Male pores: segment 15 often encroaching onto 14 and 16

Clitellum: this can start as early as 25 but usually is on segments 29 through to 34 or 35

TP: 2 humps on segments 31-33 (C)

Setae: closely paired (D)

Head: epilobic

Length: 40-180 mm. **Diameter:** 2-7 mm

Internal characters

Seminal vesicles: 4 pairs 9-12

Spermeathecae: 2 pairs 9/10 and 10/11

Califerous glands: 10-12

Aporrectodea cupulifera

(Tetry, 1937)

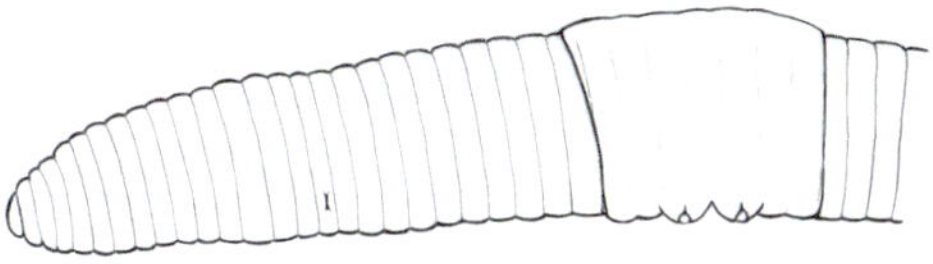

Very rare

Ecological group: endogeic

Colour in life: pink/grey occasionally very deep red head resembling *A. longa* (p. 48)

Habitat: only recorded in southern Ireland

External characters

Male pores: segment 15 A B

Clitellum: starts on 25 or usually 26 through to 32

TP: 2 humps or discs on segments 28 and 30, or span part of 27 through to 31 B

Setae: closely paired

Head: epilobic

Length: 35-50 mm

Internal characters

Calciferous glands: ½ 10-13

Seminal vesicles: 9-12

Spermathecae: 3 pairs in 9, 10 and 11

Aporrectodea icterica

(Savigny, 1826)

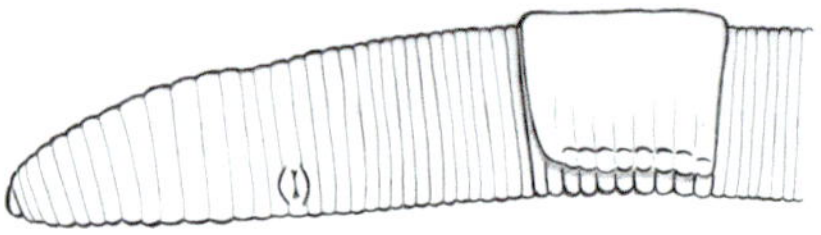

Rare

Ecological group: endogeic

Colour in life: usually grey, but can be variable; yellows, grey or sometimes with a brown tint

Habitat: found in grasslands and orchards

External characters (A) (B)

Male pores: segment 15 often encroaching onto segments 14 and 16

Clitellum: 33 or 34 through to 42-43 (rarely 44)

TP: long thin band; 35 or 36 through to 42 or 43

Setae: closely paired

Head: epilobic

Length: 50-140 mm. **Diameter:** 3-6 mm

Internal characters

Calciferous glands: ½ 10-13

Seminal vesicles: 9-12

Spermathecae: 3 pairs in 9, 10 and 11

Aporrectodea limicola

(Michaelsen, 1890)

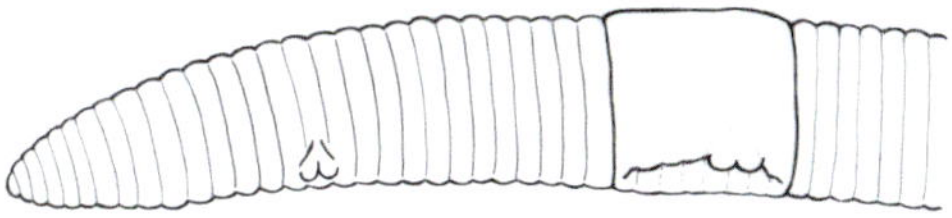

Rare with a patchy distribution (more common in the Channel Islands)

Ecological group: endogeic

Colour in life: greyish colour with pinky anterior

Habitat: waterlogged soils

External characters A B

Male pores: segment 15, encroaching onto segments 14 and 16

Clitellum: starts rarely on segment 28 normally 29-35 sometimes extends to 36

TP: 2 humps (like *A. caliginosa*) on segments 33 and 34 B

Setae: closely paired

Head: epilobic

Length: 40-100 mm. **Diameter:** 3-4 mm

Internal characters

Seminal vesicles: 4 pairs 9-12

Spermathecae: 2 pairs: 9/10 and 10/11

Aporrectodea longa

(Ude, 1885)

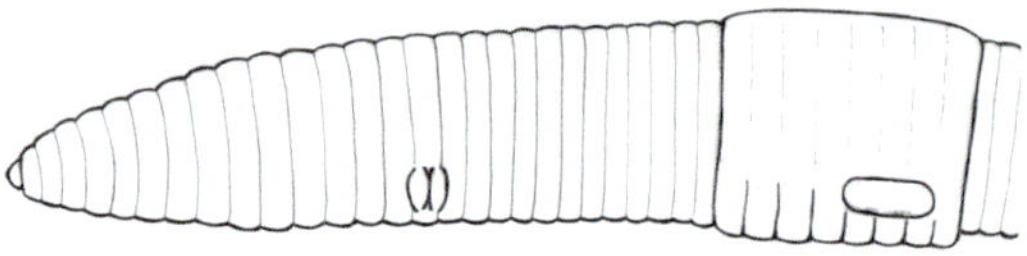

Common

Ecological group: anecic

Colour in life: distinctive black head (A)

Habitat: most abundant in grasslands and gardens. Disturbance tolerant

Despite being anecic this species also consumes soil. Sometimes produces litter middens similar to those produced by *Lumbricus terrestris*

Burrows more branched than those produced by *L. terrestris* (Andriuzzi *et al.*, 2016)

Casts: On surface

External characters (B) (C) (D)

Male pores: segment 15 occasionally encroaching onto 14 and 16

Clitellum: starts on 27 or 28 and extends through to 35 or 36

TP: a band, occasionally starting on 31 but usually 32 through to 34 (C)

Setae: closely paired

Head: epilobic (D)

Length: 90-170 mm. **Diameter:** 4-9 mm

Internal characters

Seminal vesicles: 4 pairs 9-12

Spermathecae: 2 pairs 9/10 and 10/11

Califerous glands: 10-12

Aporrectodea nocturna

(Evans 1946)

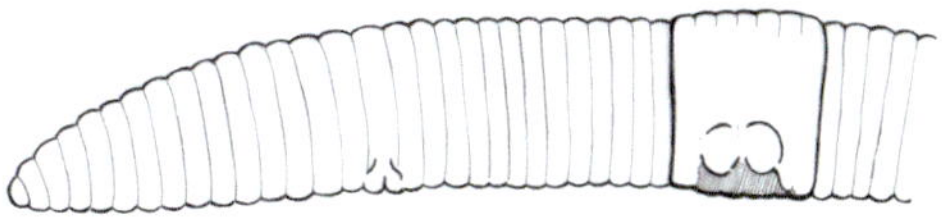

Rare

Ecological group: anecic

Colour in life: dark (red-brown) head (similar to *A. longa* p. 48)

Habitat: Disturbance-tolerant species found in gardens, pastures, forests and arable fields. Very common in grassland. Not found in the most acidic habitats such as heaths and mires

Casts: Mainly on the surface

External characters A B

Male pores: segment 15 often encroaching onto 14 and 16

Clitellum: this can start as early as 25 but usually is on segments 29 through to 34 or 35

TP: 2 humps on segments 31-33 B

Setae: closely paired

Head: epilobic

Length: 90-180 mm. **Diameter:** 2-7 mm

Internal characters

Seminal vesicles: 4 pairs 9-12

Spermeathecae: 2 pairs 9/10 and 10/11

Califerous glands: 10-12

Aporrectodea rosea

(Savigny, 1826)

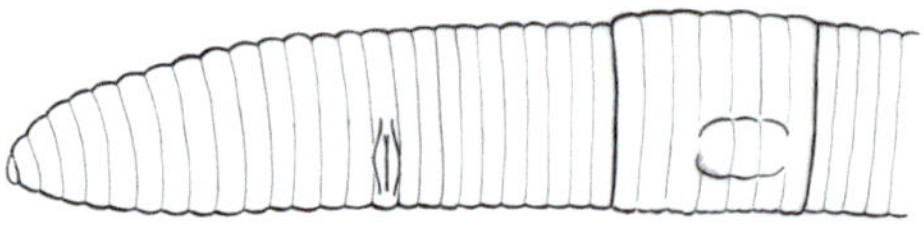

Common

Ecological group: endogeic

Colour in life: grey/pink colour, saddle often orange (A)

Habitat: universal, often found in woodland and grassland. Some studies indicate higher abundance in neutral and base-rich soils, and when soil moist. Disturbance tolerant

Casts: mainly underground but occasionally casts on the surface. Zicsi (1974) found 87.2% of casts were underground

External characters (B) (C) (D)

Male pores: usually only on segment 15, but occasionally encroaching onto 16

Clitellum: can start as early as segment 23 but usually found on 25 or 26 through to 32 or 33

TP: usually a raised bump on segments 29 to 30-31, occasionally only showing on half of segment 31 (C)

Setae: closely paired

Head: epilobic

Length: 20-110 mm. **Diameter:** 2-6 mm

Internal characters

Seminal vesicles: 4 pairs 9-12

Spermeathecae: 2 pairs 9/10 and 10/11

Califerous glands: 10-12 lateral pouches in 10

Bimastos eiseni
(Levinsen, 1884)

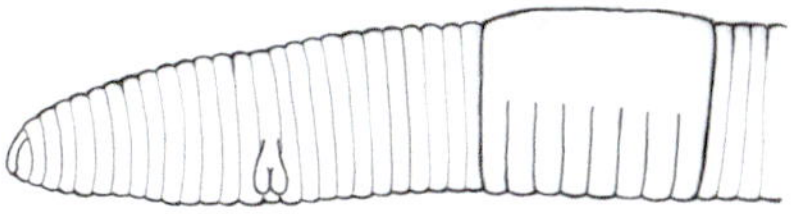

Common

Ecological group: epigeic

Colour in life: dull red or brown dorsally A

Habitat: Under bark, in tree hollows and in dead wood mainly but can also be found in some acidic soils, moorland, bogs and broadleaved woodlands

External characters B C

Male pores: segment 15

Clitellum: variable, starting anywhere from 24 but usually 27 to 32-33

TP: not present

Setae: closely paired

Head: tanylobic

Length: 30-64 mm. **Diameter:** 2-5 mm

Internal characters

Seminal vesicles: 2 pairs in 11 and 12

Spermathecae: none

Califerous glands: in 10-12

Bimastos rubidus

(Savigny, 1826)

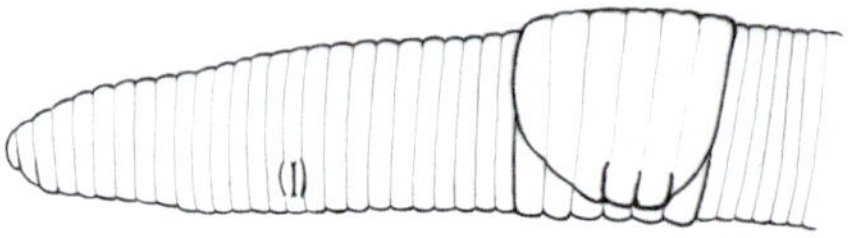

Common

Ecological group: epigeic

Colour in life: deep red on dorsal surface A

Habitat: often found under the bark of rotting wood, and under stones or dung. Favours soils with high organic content and acidic soils. Can be found in wood ant nests

Casts: On soil surface

External characters B C D E

Male pores: segment 15, encroaching onto segments 14 and 16 in some cases

Clitellum: usually 26 or 27 (very rarely starts on 25) through to (rarely 31) 32-33

TP: a swelling, looks like 4 squares with 2 on each side usually 29-30 but can also start on 28 or finish on 31 C

Occasionally TP is completely absent D

Setae: widely paired E

Head: epilobic

Length: 20-100 mm. **Diameter:** 2-6 mm

Internal characters

Seminal vesicles: 3 pairs 9, 11, 12

Spermeathecae: 2 pairs 9/10 and 10/11

Califerous glands: with lateral diverticular in 11

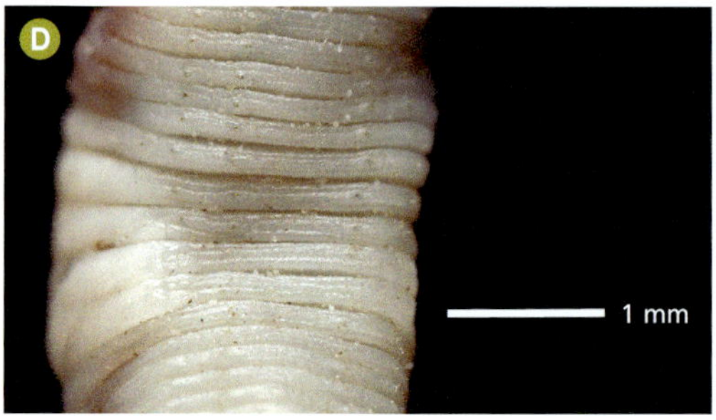

Dendrobaena attemsi

(Michaelsen, 1902)

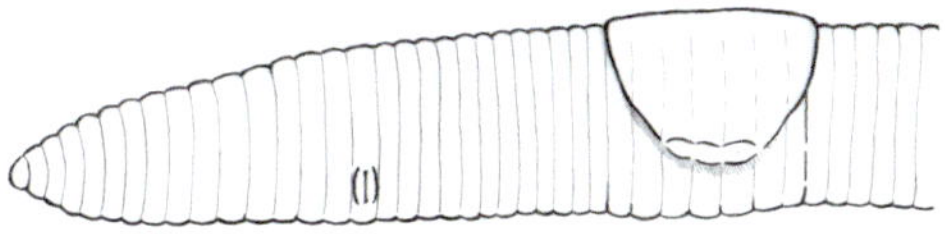

Uncommon

Ecological group: epigeic

Colour in life: darkish red dorsally or pale.

Once preserved loses colour very quickly A

Habitat: Found under bark, in leaf litter, rotting wood and in the soil (often peaty soils or in greenhouses). Prefers acidic conditions, and can be locally common in acid woodlands

External characters B C

Male pores: segment 15 very occasionally encroaching onto 16

Clitellum: 28 or usually 29 to 34

TP: raised band on 30-32 C

Setae: widely paired

Head: epilobic

Length: 26-70 mm. **Diameter:** 1-3 mm

Internal characters

Seminal vesicles: 4 pairs 9-12

Spermeathecae: 2 pairs 9/10 and 10/11

Dendrobaena hortensis

(Michaelsen, 1890)

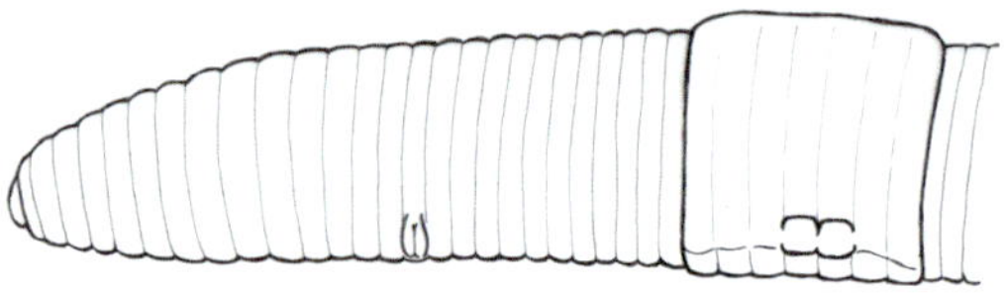

Rare

Ecological group: epigeic

Colour in life: variable colour; reddish generally at front end turning paler to white A

Habitat: mainly found in broadleaved woodland on acidic soils, frequent in compost; much more abundant in southern Europe

External characters B C

Male pores: segment 15 B

Clitellum: occasionally starts on 26 usually 27-33 C

TP: a swelling on 30 and 31, occasionally starting on 29 and extending onto 32 C

Setae: widely paired

Head: epilobic

Length: 20-50 mm. **Diameter:** 1.5-5 mm

Fecundity: Cocoons can contain 1-4 juveniles. Can produce on average 4.7 juveniles per week. Incubation time on average 4.5 days, but can be as little as 3.5 days to reach maturity (Csuzdi, 1988)

Internal characters

Seminal vesicles: 3 pairs (ocassionally 2) 9, 11, 12 (occasionally just 11, 12)

Spermeathecae: 2 pairs 9/10 and 10/11

Califerous lamellae: 11-13

Last pair of hearts: 11

Dendrobaena octaedra
(Savigny, 1826)

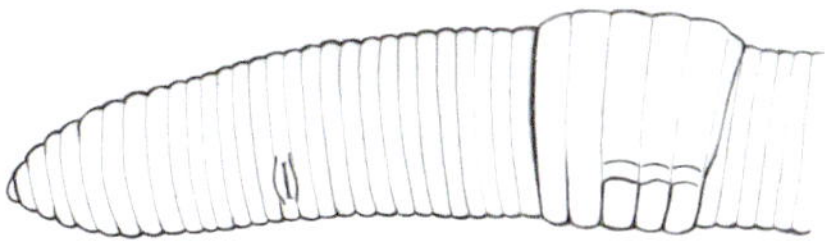

Uncommon (but can be locally abundant)

Ecological group: epigeic

Colour in life: dark red between head and clitellum dorsally, pale ventrally (A)

Habitat: Under litter or bark or stones. Prefers broadleaved woodland on acidic soils

Casts: exclusively on the soil surface

Mainly parthenogenetic. The cocoons are frost resistant, so can survive low temperatures

External characters (B) (C) (D) (E)

Male pores: segment 15

Clitellum: 27 or 28 but usually 29 through to 33 or 34

TP: usually a distinctive thin band (C), but sometimes a raised area; 31-33 (D)

Setae: widely paired (can be distinctive) (E)

Head: epilobic

Length: 20-60 mm. **Diameter:** 2-5 mm

Internal characters

Seminal vesicles: 3 pairs 9,11,12

Spermeathecae: 3 pairs 9/10, 10/11, 11/12

Califerous glands: 10-12

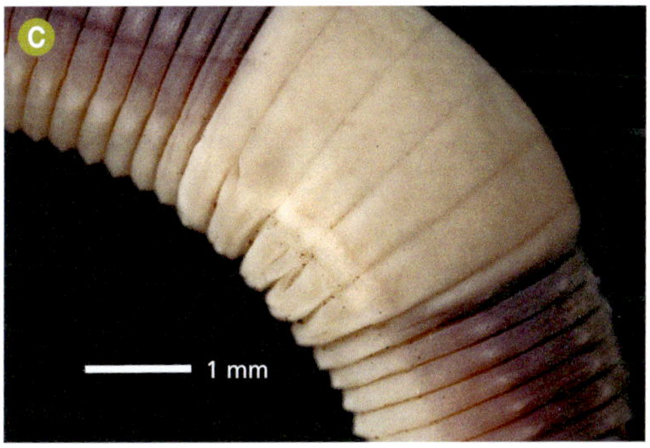

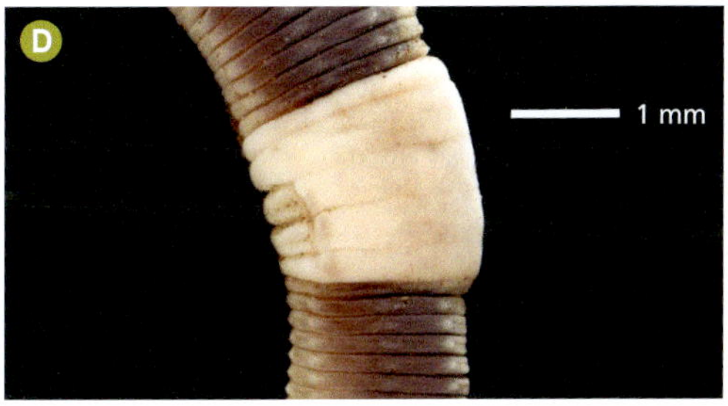

Dendrobaena pygmaea
(Savigny, 1826)

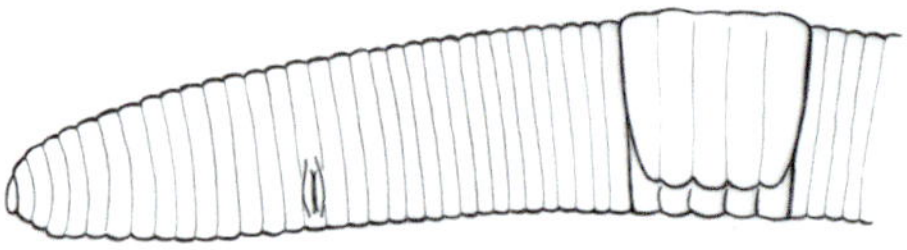

Very rare

Ecological group: epigeic

Colour in life: variable, some individuals light in colour with a reddish or greyish tint, others unpigmented (A)

Habitat: In rotting wood, mossy banks of streams, well drained soils and some broadleaved woodland

Often parthenogenetic

External characters (B)

Male pores: segment 15

Clitellum: 32 or 33 through to 37-38 (B)

TP: usually not present. If present, a ridge in 35-37

Setae: widely paired

Head: epilobic

Length: 15-45 mm. **Diameter:** 0.5-1.2 mm

Internal characters

Seminal vesicles: 2 pairs in 11 and 12, but sometimes absent

Dendrobaena veneta
(Rosa, 1886)

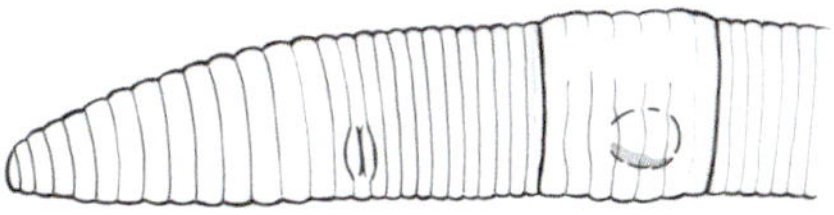

Common

Ecological group: epigeic.

Colour in life: deep red, a stripy appearance when stretched A

Habitat: well drained soils, some broadleaved woodlands. Used in vermicomposting

Fecundity: cocoons produce 1-2 juveniles. Average 1.4 juveniles per week (Edwards, 1998). Incubation time 21-31 days, 82-99 days to maturity (Zicsi, 1985)

Life expectancy of the species: 3.2 years (Zicsi, 1974)

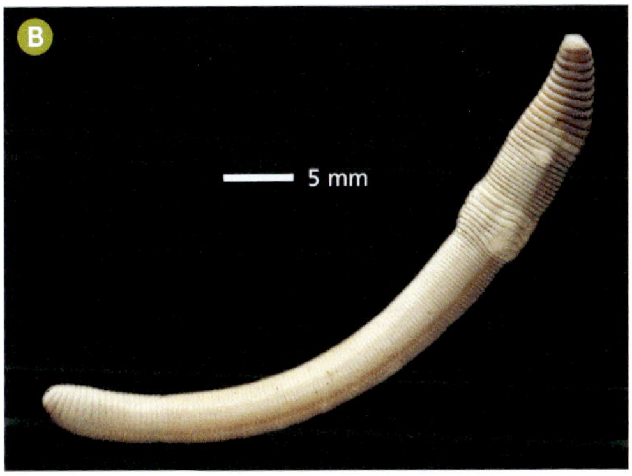

External characters B C D

Male pores: segment 15

Clitellum: 26-27 through to 32 or 33

TP: a swelling; usually 30-31 but can start on 29 and extend to 32 C

Setae: widely paired D

Head: epilobic (but can be tanylobic)

Length: 30-155 mm. **Diameter:** 0.5-1.2 mm

Internal characters

Seminal vesicles: 4 pairs 9-12 (occasionally 3 or even 2)

Spermeathecae: 2 pairs 9/10 and 10/11

Califerous lamellae: 1/2 10-11

Eisenia fetida

(Savigny, 1826)

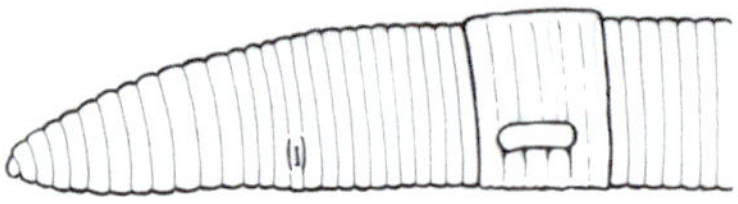

Common (in compost)

Ecological group: epigeic

Colour in life: dark dorsally, appears striped when moving. Produces a bright yellow fluid over its body surface if disturbed (see inset), which to some people has a strong odour. *Eisenia andrei* is closely related but is not striped A

Habitat: places with very high organic matter such as compost bins and under dung. Used in vermicomposting; known as the tiger worm

Fecundity: 70-98 days to maturity (Zicsi, 1985)

Life expectancy of the species: 2.25 years (Zicsi, 1974), maximum 4.5 years (Lee, 1985). Reproductive rate of 0.644 juveniles per adult per day (Csuzdi & Zicsi, 1988)

External characters B C D

Male pores: segment 15

Clitellum: starting on 24-27 through to 32, sometimes onto 33 or even 34

TP: thin bands running from 27 or mainly 28 through to 30 or 31, occasionally onto 32 C D

Setae: closely paired

Head: epilobic

Length: 26-130 mm. **Diameter:** 2-6 mm

Internal characters

Seminal vesicles: 4 pairs 9-12

Spermathecae: 2 pairs 9/10 and 10/11

Califerous glands: 10-12

Eiseniella tetraedra
(Savigny, 1826)

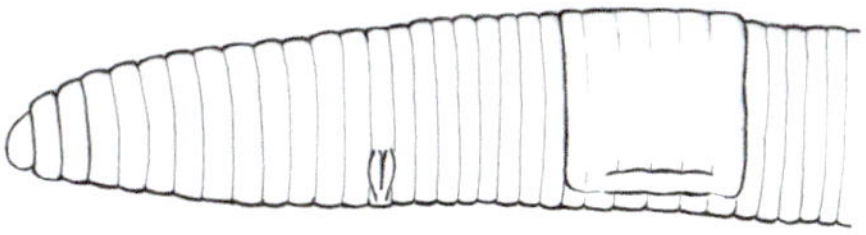

Common in wet habitats

Ecological group: endogeic

Colour in life: pale worm with green-brown or pink/red appearance A

Habitat: waterlogged soils

External characters B C D

Male pores: usually on 13 (very rarely on 11, 12, 14 or 15) Different parthenogenetic morphs with differing male pore positionings D

Tail: a 'square' shape

Clitellum: 22-23 to 26-27

TP: long thin bands on 23-26 occasionally extending onto 22 and 27 C

Setae: closely paired

Head: epilobic

Length: 20-80 mm. **Diameter:** 1.5-4 mm

Internal characters

Seminal vesicles: 4 pairs 9-12

Spermeathecae: 2 pairs 9/10 and 10/11

Califerous glands: with diverticular in 10

Helodrilus oculatus
Hoffmeister, 1845

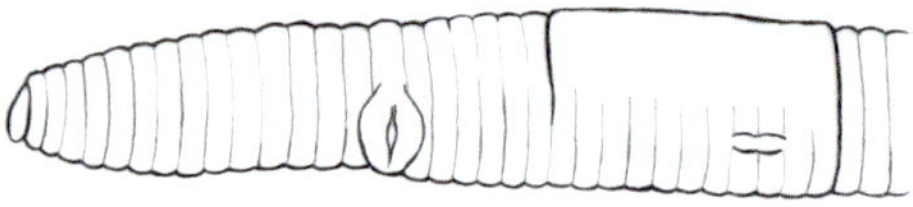

Very rare

Ecological group: endogeic

Colour in life: pale worm occasionally with some black markings, very 'stringy' looking A

Habitat: only found in waterlogged soils, especially in broadleaved woodlands. Alkaline conditions preferred. Often populations of just juveniles are found

External characters B C D E

Male pores: 15, but extending onto 14 and 16, often very large D

Clitellum: usually 22 (occasionally 21) through to 32

TP: ridge on 29 and 30 C

Setae: closely paired E

Head: epilobic

Length: 35-80 mm. **Diameter:** 1-2 mm

Internal characters

Seminal vesicles: 2 pairs 11 and 12

Olaf Schmidt

Kenleenus armadus

Blakemore, 2012

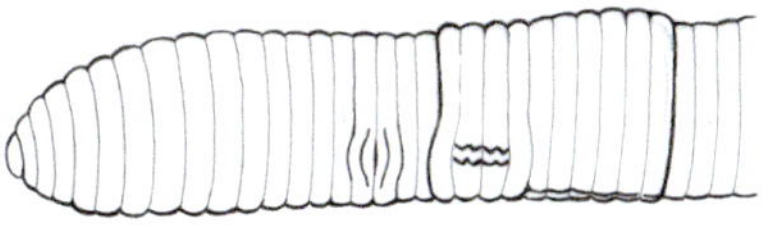

Very rare

Ecological group: endogeic

Colour in life: pale rosy pink

Habitat: Specimen found on farmland in Dublin

External characters B C

Male pores: segment 15 extending onto segments 14 and 16 (very large)

Clitellum: 19-28 (29)

TP: flared, mostly 20-21, but extends into 19 and/or 22

Setae: widely paired

Head: epilobic

1st dorsal pore: 7/8

Length: 35-65 mm. **Diameter:** 1-2 mm

Discovered in Ireland by Olaf Schmidt in 2012. Initially thought to be the French species *Prosellodrilus amplisetosus* (moved to the genus *Kenleenus* by Blakemore, 2012).

Blakemore described it as a new species due to molecular evidence as well as positioning of the 1st dorsal pore (7/8 for *K. armadas*, 5/6 or 6/7 for *amplisetosus*) prostomium shape (epilobic closed for *K. armadas*, prolobic for *P. amplisetosus* and genital papillae.

Internal characters

Seminal vesicles: 2 pairs occasionally 10 and 11

Spermathecae: 2 pairs 14 and 15

Lumbricus castaneus
(Savigny, 1826)

Common

Ecological group: epigeic

Colour in life: deep red anteriorly A

Habitat: Usually found in rotting logs and under litter of forests. Prefers soils rich in organic matter, although can be found in most habitats. Often found in grassland, woodland and hedgerows

Casts: 80% on soil surface (Zicsi, 1974)

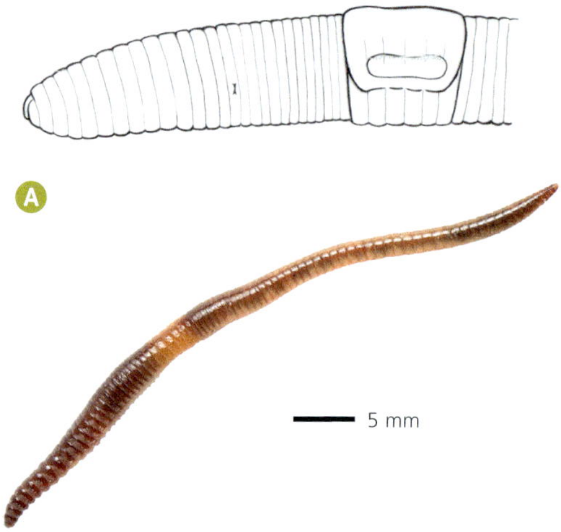

External characters B C

Male pores: segment 15 (very small)

Clitellum: 28-33 but sometimes encroaches onto 27 and 34

TP: bands; 29-32 C

Setae: closely paired

Head: tanylobic

1st dorsal pore: 6/7

Length: 30-85 mm. **Diameter:** 2-5 mm

Internal characters

Seminal vesicles: 4 pairs 9-12

Spermeathecae: 2 pairs 9/10 and 10/11

Califerous glands: 10-12, lateral pouches in 10

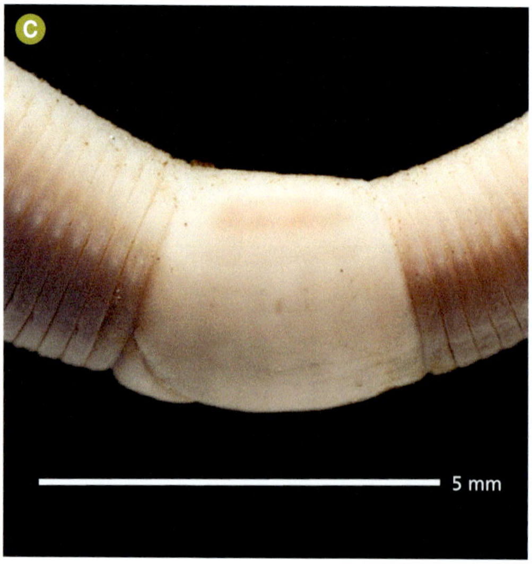

Lumbricus festivus

(Savigny, 1826)

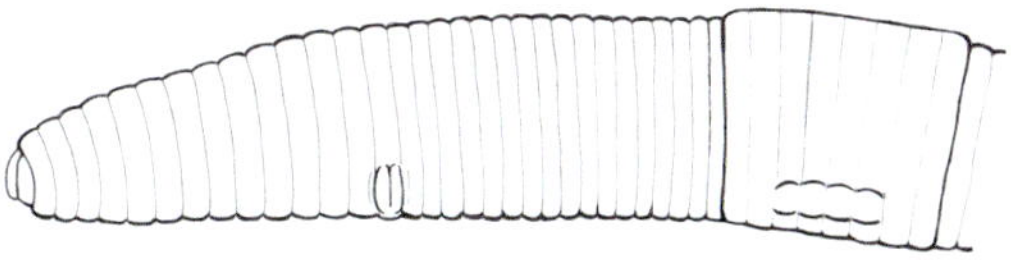

Rare

Ecological group: epigeic

Colour in life: deep red anteriorly A

Habitat: prefers grasslands and arable fields. Mainly damp often acidic soils, can also be found under logs and on river banks

External characters B C D E

Male pores: segment 15

Clitellum: starting on 33 or 34 through to 40

TP: bands; 35-38 C D

Setae: closely paired

Head: tanylobic E

1st dorsal pore: 5/6

Length: 48-110mm. **Diameter:** 3.5-6 mm

Internal characters

Seminal vesicles: 3 pairs 9,11,12

Lumbricus friendi

Cognetti, 1904

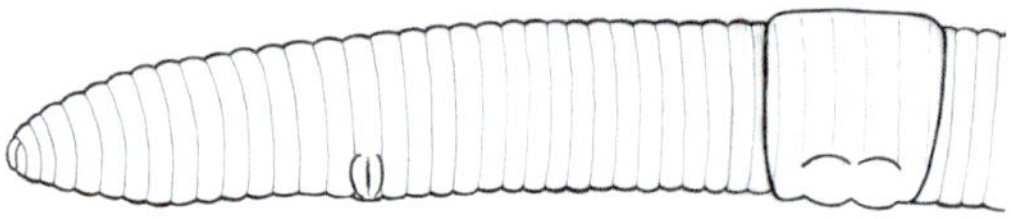

Very rare

Ecological group: anecic

Colour in life: purple-red dorsally

Habitat: very moist conditions, woodlands and grasslands. Calcareous soils. Few records in the UK or Ireland. Produces near vertical burrows and middens similar to *Lumbricus terrestris*

Fecundity: Hatching in perfect conditions approximately 92 days (66-136); most cocoons produce one juvenile. Hatchlings reach maturity in 20 days. remain low. Cocoon production per week 2.16 per individual, cocoon mass approximately 28g

External characters A B

Male pores: 15, but occasionally encroaching slightly onto 14 and 16 A

Clitellum: half 32-33 through to 37-38

TP: thin ridge; 34-36 B

Setae: closely paired

Head: tanylobic

Length: 90-200 mm. **Diameter:** 4-8 mm

Internal characters

Seminal vesicles: 3 pairs 9,11,12

Spermeathecae: 2 pairs 9/10 and 10/11

Califerous glands: 10-12

Lumbricus rubellus

Hoffmeister, 1845

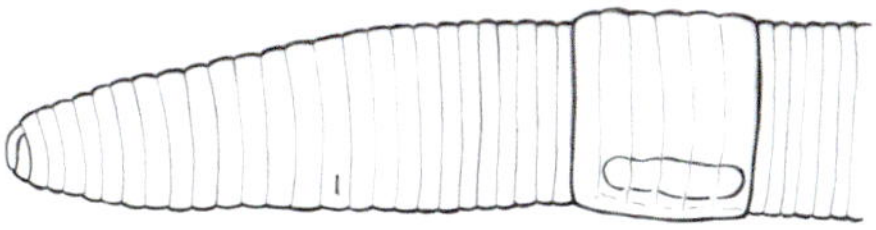

Very common

Ecological group: epigeic (although some large forms thought to act more as an anecic)

Colour in life: purple-red dorsally A

Habitat: Recorded from all habitats, particularly prevalent in organic-rich environments such as under dung as well as in acid grassland and woodland. However, it seems to have the lowest habitat specificity of all British and Irish earthworms

Casts: Zicsi (1974) reports 80% of casts on the surface

External characters B C D

Male pores: segment 15 (small)

Clitellum: 26 or 27 through to 32

TP: thick band; occasional part of segment 27 but mainly 28-31 C

Setae: closely paired

Head: tanylobic D

1st dorsal pore: 7/8

Length: Very large range: 20-150 mm

Diameter: 3-6 mm

Internal characters

Seminal vesicles: 3 pairs 9, 11, 12

Spermeathecae: 2 pairs 9/10 and 10/11

Califerous glands: 10-12, lateral pouches in 10

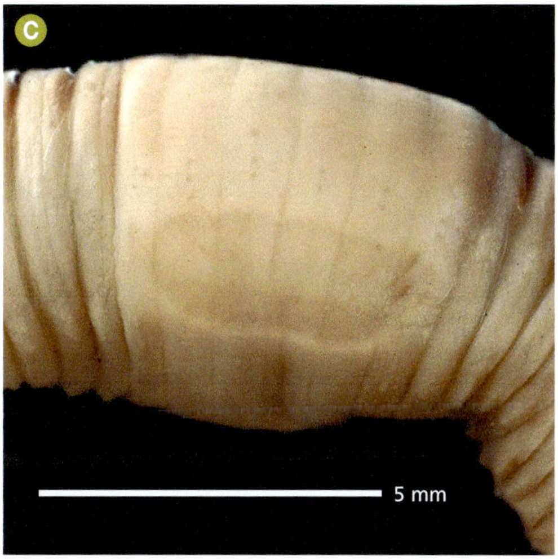

Lumbricus terrestris

Linnaeus, 1758

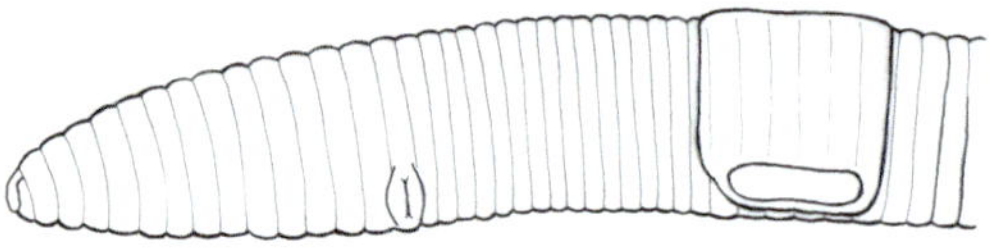

Common

Ecological group: anecic

Colour in life: deep purple-red dorsally (A)

Habitat: mainly found in grasslands and lawns, especially when undisturbed. Preference for clay, and also alkaline conditions

Casts: Mainly casts on the surface

Fecundity: usually 1 individual per cocoon. Adult produces 0.72 cocoons per week. Incubation time 90 days (Lowe & Butt, 2005)

External characters (B) (C)

Male pores: segment 15 often encroaching onto segments 14 and 16

Clitellum: 31 or 32 through to 37

TP: a thick ridge, can be shaped like a canoe; 32 or 33 through to 36 or 37

Setae: closely paired

Head: tanylobic

1st dorsal pore: 7/8

Length: 90-350 mm. **Diameter:** 6-10 mm

Internal characters

Seminal vesicles: 3 pairs 9, 11, 12

Spermeathecae: 2 pairs 9/10 and 10/11

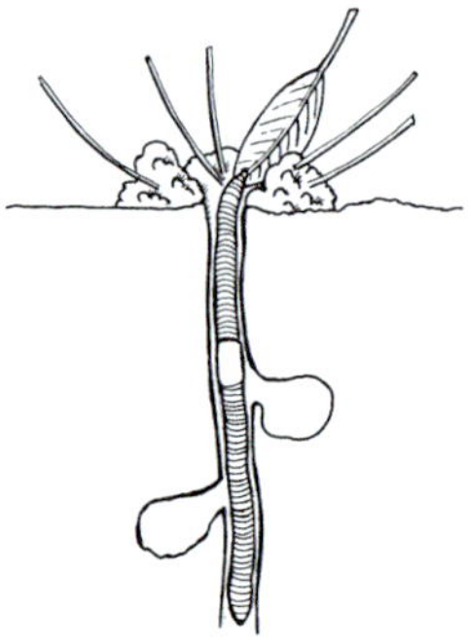

L. terrestris produces vertical burrows with little or no branching, except for occasional side branches open to the surface (Nuutinen & Butt, 2003). Burrows are permanent structures which can be inherited by a juvenile earthworm on the owner's death (Grigoropoulou *et al.*, 2009). They can be 1-2.5 metres deep (Edwards & Bohlen, 1996). Adults can be found resting about 20-25 cm deep, but at much greater depths during during aestrus.

A midden is constructed where the burrow opens to the surface; a small mound of leaf litter, twigs and casts (see Fig. 3, p. 6). In well populated areas, these middens are generally about 13 cm apart and can form a regular pattern (Grigoropoulou & Butt, 2010).

Microscolex phosphoreus
(Dugès, 1837)

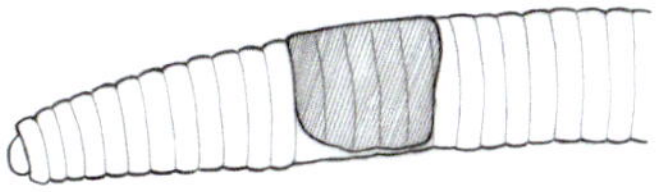

Very rare

Ecological group: endogeic

Colour in life: can be bioluminescent at night

Habitat: found in areas of high organic matter such as compost heaps and also found on golf courses

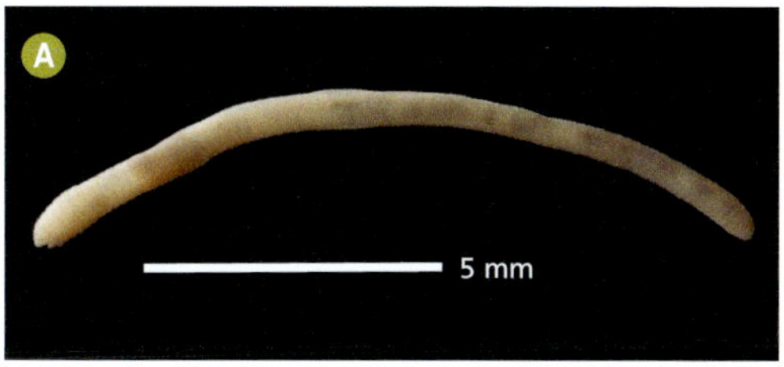

External characters A B

Male pores: segment 17

Clitellum: 13-17 and a very different shape to that of the Lumbricidae family of which most of the British and Irish fauna is part

TP: not present but male and prosthetic pores on 17 and female pores often visible on 14

Setae: patterns different

Head: epilobic

Length: 10-35 mm. **Diameter:** 1-1.5 mm

Internal characters

Spermeathecae: 1 pair 8/9

Gizzard: in 5

Murchieona muldali

(Omodeo, 1956)

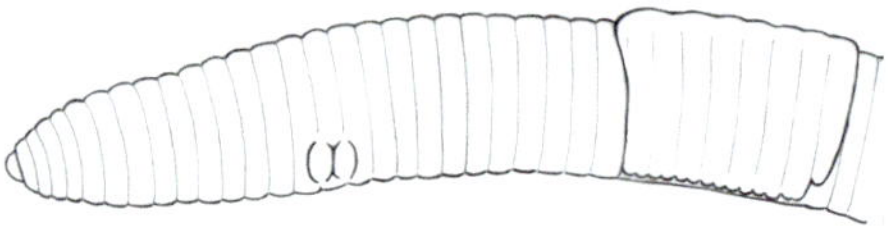

Rare

Ecological group: endogeic

Colour in life: pink-grey

Habitat: favours areas with a high moisture content and pH. Can be found in woodland and grassland; particularly abundant in field margins

Fecundity: Never more than one embryo in each cocoon. Populations contain a mix of individuals produced by sexual reproduction and those produced through parthenogenesis

External characters A B

Male pores: segment 15 encroaching onto segments 14 and 16

Clitellum: 26-27 to 33-34 A B

TP: unlikely to be present

Setae: closely paired

Head: epilobic

Length: 20-60 mm. **Diameter:** 1.2-2 mm

Internal characters

Seminal vesicles: 2 pairs 11-12, although often absent

Octolasion cyaneum

(Savigny, 1826)

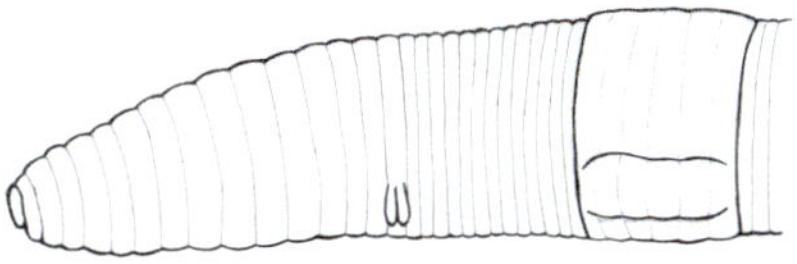

Common

Ecological group: endogeic

Colour in life: no pigmentation or a light blue-grey often with a yellow tip to the tail A

Habitat: prefers wet conditions but has a wide tolerance (Reynolds, 1977) and can be found in most habitats

Fecundity: Mainly parthenogenetic. 1 or 2 juveniles per cocoon, 20% twins, 128-129 day incubation. 24-44 weeks to maturity (Lowe *et al.*, 2008)

External characters B C

Male pores: segment 15

Clitellum: 29-34

TP: white bands on 30-33, or can start on 29 and extend to 34, covering the entire length of the clitellum

Setae: very small and difficult to see often more closely paired at the anterior but widely paired in the posterior region

Head: epilobic C

Length: 40-180 mm. **Diameter:** 5-8 mm

Internal characters

Seminal vesicles: 4 pairs 9-12

Spermeathecae: 2 pairs 9/10 and 10/11

Califerous glands: 10-12

Last pair of hearts: 11

Octalasion lacteum

(Örley, 1881)

Common

Ecological group: endogeic

Colour in life: variable, but usually greyish. May have a yellow tail (A)

Habitat: preference for very moist habitats

Casts: Mainly in the soil

Usually parthenogenetic

External characters (B) (C) (D)

Male pores: segment 15 encroaching onto segments 14 and 16

Clitellum: 30-35

TP: a band; half 30, 31-34, half 35; may cover the entire length of the clitellum (C)

Setae: more closely paired anteriorally but widely paired posteriorly (D)

Head: epilobic

Length: 25-160 mm. **Diameter:** 2.5-6 mm

Internal characters

Seminal vesicles: 4 pairs 9-12

Spermeathecae: 2 pairs 9/10 and 10/11

Califerous glands: 10-12

Last pair of hearts: 11

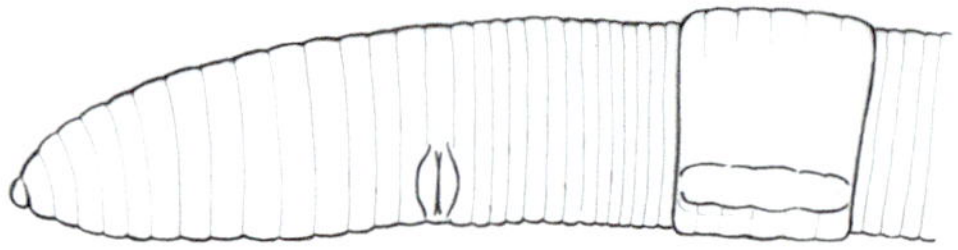

Satchellius mammalis

(Savigny, 1826)

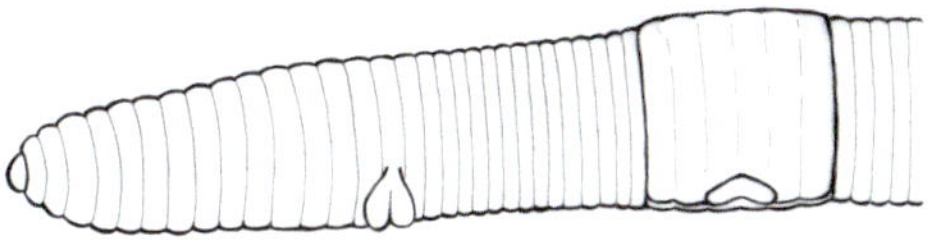

Uncommon

Ecological group: epigeic

Colour in life: deep red in life A

Habitat: found in woodlands, areas of high organic matter and on river banks. Favours base-rich conditions

External characters B C

Male pores: 15 encroaching onto 14 and 16

Clitellum: 31-36

TP: a semi-circular raised band; 33 and 34 B C

Setae: widely paired

Head: epilobic

Length: 24-41 mm. **Diameter:** 1.5-3 mm

Internal characters

Seminal vesicles: 4 pairs 9-12

Spermeathecae: 2 pairs 9/10 and 10/11

Sparganophilus tamesis

Benham, 1892

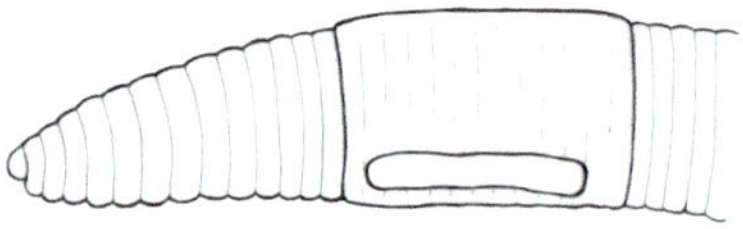

Very rare

Ecological group: endogeic; although does not appear to conform particularly well to the groupings. The author of the species believes they stay deep in wet mud for much of the year coming to the surface among the roots of plants during the summer months for reproduction

Colour in life: pale pink

Habitat: found in saturated soils and river beds. The most aquatic of the earthworms of the UK and Ireland

External characters B C

Male pores: 19 (inconspicuous)

Clitellum: 14 or 15 through to 26

TP: shape variable; length variable: present along most of the clitellum, always 19-22, usually 15-24 B

Setae: closely paired C

Head: epilobic

Length: 70-200 mm. **Diameter:** 1-5 mm

Spermathecal pores: paired

Internal characters

Spermeathecae: 3 pairs 6/7 7/8 8/9

Prostate glands: present

Earthworms found in artifical environments

Sometimes exotic invertebrates can survive in isolated pockets within artificially warm and humid environments. This section contains a key to the 8 exotic species that have been recorded in greenhouses in the UK and Ireland within the last 20 years. The only exception is *Dichogaster bolaui*, now regularly found in Europe and likely to turn up here. Other exotic species have been recorded in the more distant past, particularly at Kew (Table 1 part 2, p. 3). Few active earthworm recorders currently collect in greenhouses. I hope this guide will encourage those biological recorders who monitor greenhouses for natural history societies, as well as greenhouse managers and their teams, to add earthworms to the fauna they look out for.

External characters

Spermathecal pores

The external pore openings of the spermathecae can be diagnostic. They are usually situated between setae. This key only asks whether the pores are dorsal; a very unusual position for them. They are more readily visible on fresh specimens where darker head pigment is present

Male and prostatic pores

In the Lumbricidae the male pores are located anterior to the clitellum and on segments 13, 14 or 15. But in other families the male pores are often posterior to the clitellum, sometimes paired with the prostatic pores (some families of earthworms possess prostates). The genital markings around these pores are a useful character, taking a range of sizes and shapes including circles.

Female pore

Although the female pore is rarely diagnostic, its position is useful to check segment numbers. In Megascolecidae the female pore is a small whitish mark on the clitellum, segment 14.

Clitellum

A familar character for British earthworms which can look different in exotic families. In Fig. 57 it is on segments 14-16.

Figure 56. Spermathecal pores.

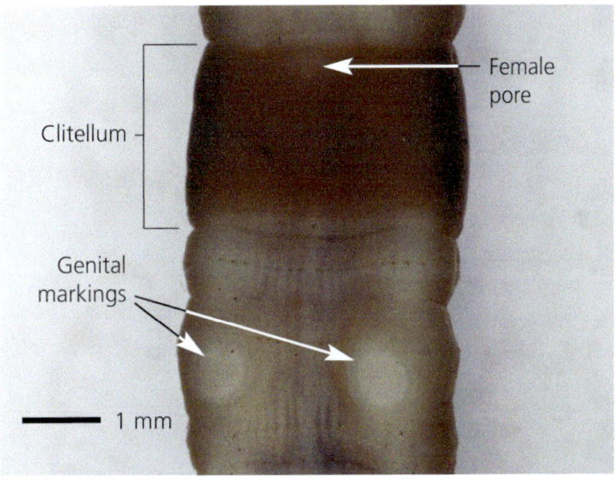

Figure 57. Clitellum and male and female pores.

Internal structures

Spermathecae

Spermathecae are sacs used to store the sperm received from another worm, often with diverticula leading from each sac. Useful characters include the size and shape of the sacs and diverticula, the attachment of the diverticula to the sacs, plus the location of where the spermathecae attach to the body wall.

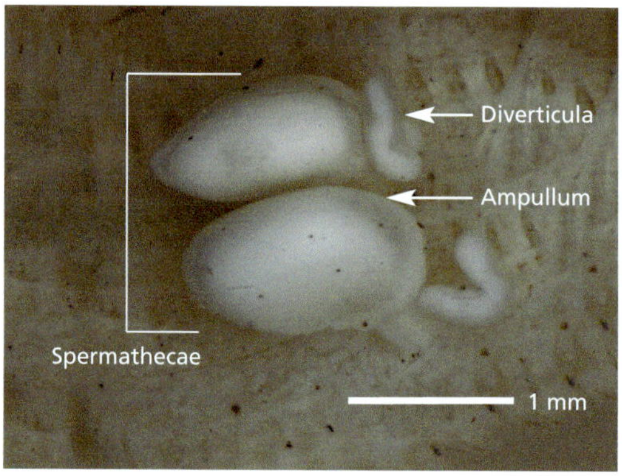

Figure 58. Spermathecae.

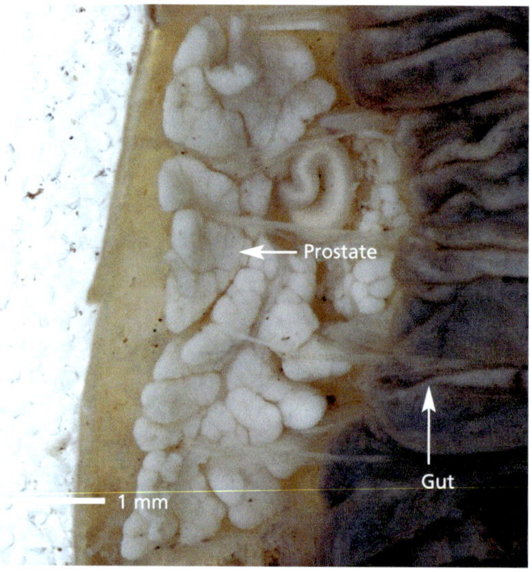

Figure 59. Prostate and gut.

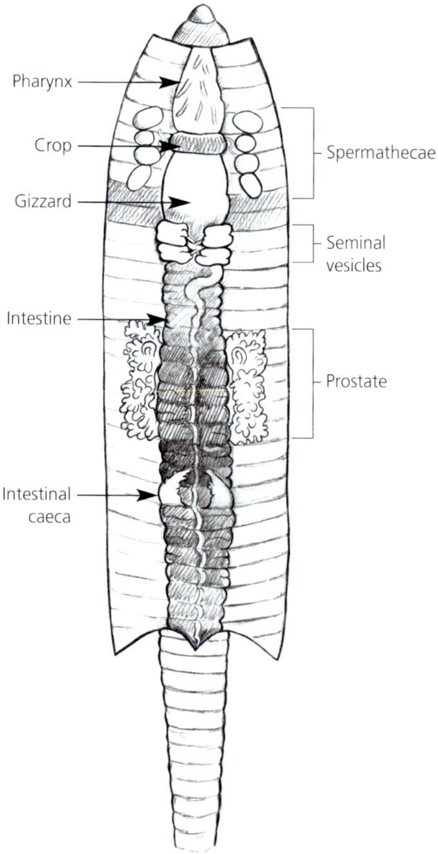

Figure 60. Internal structures in an earthworm revealed by dissection.

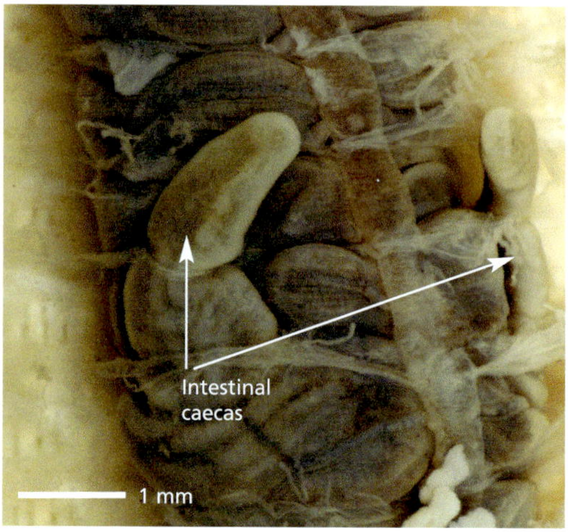

Figure 61. Intestinal caecas.

Earthworm dissection

Materials needed

- Microscope
- Dissection scissors (medical scissors designed for eye surgery are best, although expensive)
- A tray lined with wax, silcon rubber or a Tupperware pot with plastozote secured within (this needs to fit under the microscope)
- Entomological pins
- Water that has been sitting for some time or alcohol
- Forceps
- Squeezy pipette

Figure 62. A pinned specimen.

Dissection technique

If the worm is large enough to hold and cut, hold it in your left hand (if right handed) with the dorsal surface of the worm towards you. Using dissection scissors, make an incision two thirds down the length of the worm from the head, taking care only to puncture the cucticle and not the gut beneath. Carefully cut through the cuticle following a line just to the side of the dorsal pores and blood vessel. Try not to destroy the pores, in case you need to check them later. Once segment 3 is reached, place two transverse cuts to help ease the opening of the worm, then go back to your starting point and do the same. The worm should now be ready to be pinned.

If your worm is too small to hold and cut, place the worm into the dissection dish and pin the very tip of the head down into the dish and then place another pin near the tail. Use forceps to lift the cuticle of the worm, then make an incision about two thirds along, which should be a good 10-20 segments after the clitellum. Take care only to puncture the cucticle of the worm and not the gut beneath. Carefully cut all the way to the head of the worm but stop about 3 segments from the very tip and make small transverse snips as detailed above. Return to the place where you made the first incision and make a sideways cut there too.

If the worm is not yet anchored to the tray, stick a pin in the first segment to hold it in place. Then carefully open the worm starting at the posterior end where the septa are weaker. Avoid using too many pins, as this can obsure your view later, and try to place each pin at a slight angle. Towards the head end the septa between each segment can be very tough, particularly for species that live in clay soils, so you may need to snip a few of these to be able to pin out the worm. Cut these close to the gut so as not to damage organs attached to the body wall. Then pin out the worm putting the last pins in segment 5 if you can, as this helps with counting segments. Take care with this pin and the couple before as delicate spermathecae may be positioned close to the cut and certainly close to where you might hope to position the pin.

Cover the worm with alcohol or water. If the gut has been punctured, there will be soil obscuring the organs. Use the squeezy pipette to push any murky water away, by sucking up clean water and blowing it over the worm to clear. The worm is now ready for Questions 4-6 in the key.

Key to earthworms found in artificial environments

Questions 1-3 can be answered using external characters, Questions 4-6 require dissection Free-living British and Irish species may also be present in greenhouses. Try the main key (pages 28-39) first if the earthworm does not immediately look different from a common species. As it is possible that further exotic species not included in this key could be found, make sure that the worm does conform to the description before making a definite record.

1. Does the clitellum span segments 14 to 16?

– Yes .... go to Question 3

– No .... go to Question 2

2. Which segments does the clitellum span?

– Segments 15/16 to 22/23? .... *Pontoscolex corethurus* (p. 82)

– Segments 13/14 to 18/19 AND there is this marking on the ventral surface of the clitellum? ... *Dichogaster bolaui* (p. 80)

3. Does the worm have dorsal spermathecal pores?

– Yes .... *Amynthas rodericensis* (p. 79)

– No .... go to Question 4

Open the worm (see instructions on page 75)

4. Does the worm have intestinal caecas?

– Yes .... Go to Question 5

– No .... *Polypheretima taprobanae* (p. 82)

5. How many pairs of spermathecae are there?

Read all 4 options

- 5 pairs .... *Pithemera bictincta* (p. 81)
- 4 pairs .... Go to Question 6
- 3 pairs .... *Amynthas gracilis* (p. 78)
- 2 pairs .... *Metaphire californica* (p. 80)

6. Are the diverticular twisted (see picture, right)?

- Yes .... *Amynthas rodericensis* (p. 79)
- No .... *Amynthas corticis* (p. 78)

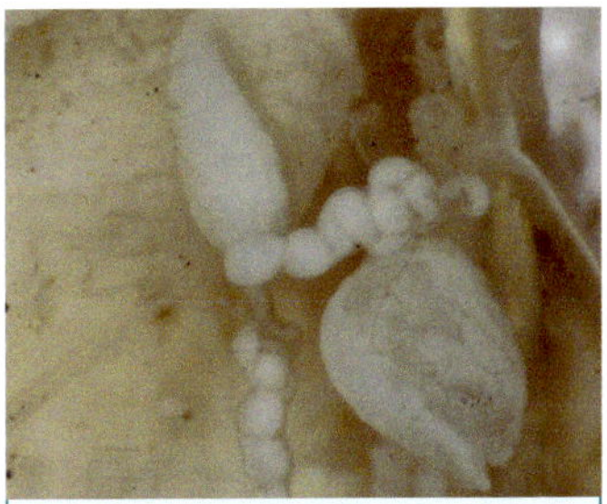

Check: Do the diverticular look like this? If so it is probably *Amynthas rodericensis*. Re-check the position of the spermathecal pores.

Species accounts for earthworms found in artificial environments

Amynthas corticis
(Kinberg, 1867)

Main characters

Clitellum: 14-16 (A)

Female pore: 14

Male pore: 18

Prostates: present

Intestinal caeca: present

Spermathecae: 4 pairs

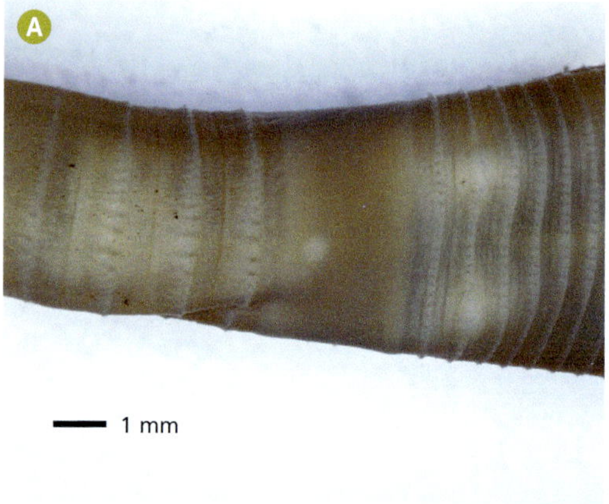

Amynthas gracilis
(Kinberg, 1867)

Main characters

Clitellum: 14-16 (B)

Female pore: 14

Male pore: 18

Prostates: present (C)

Intestinal caeca: present (D)

Spermathecae: 3 pairs

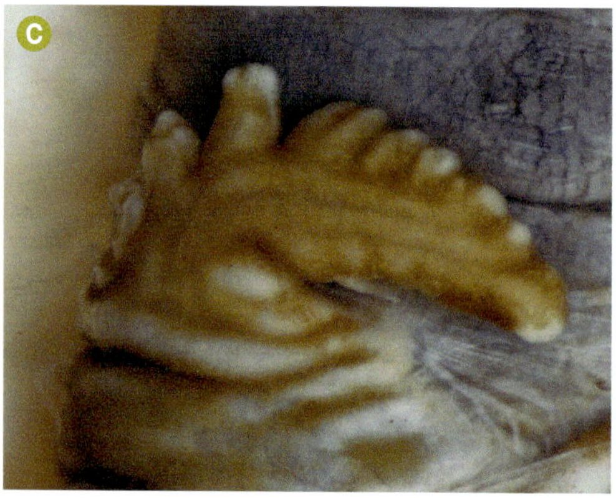

Amynthas rodericensis

(Grube, 1879)

Currently the most commonly found worm in greenhouses in the UK

Distinctive characters

- Dorsal spermathecal pores B
- 1 pair or 2 pairs (occasionally one pair and one single) of large circular glandular markings C D

Main characters

Clitellum: 14-16

Female pore: 14

Male pore: 18

Prostates: present

Intestinal caeca: present E

Spermathecae: 4 pairs: 5/6 6/7 7/8 8/9

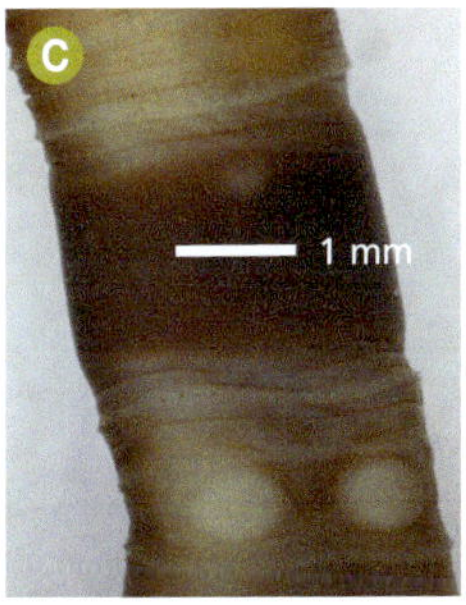

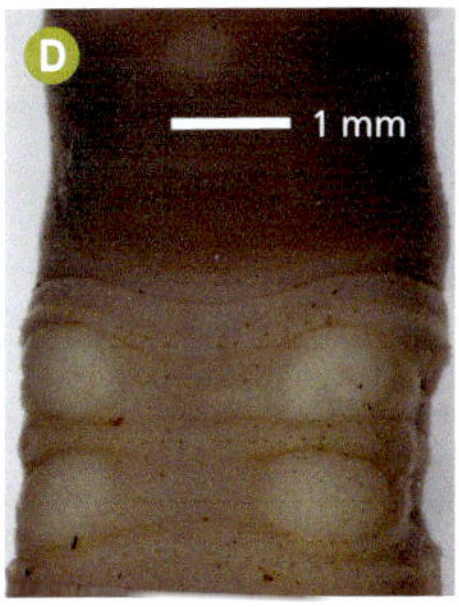

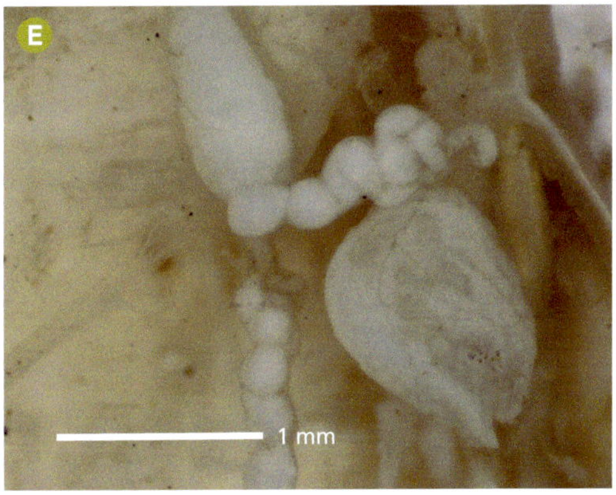

Dichogaster bolaui

(Michaelson, 1891)

Distinctive characters

- Very small (2-4 cm)
- Grooves on the inside of the clitellum between male and prostatic pores B

Main characters

Clitellum: (13)14-18 (19)

Grooves between prostatic and male pore on segments 17 and 18 B

Length: 2-4 cm

Metaphire californica

(Kinberg, 1867)

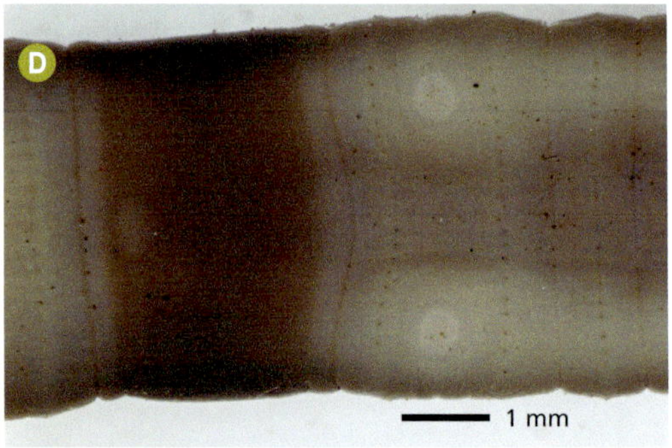

Main characters

Clitellum: 14-16 D

Female pore: 14

Male pore: 18

Prostates: present

Intestinal caeca: present E

Spermathecae: 2 pairs in segments 8 and 9

Pithemera bictincta
(Perrier, 1875)

Distinctive characters

- Bean-shaped glanular swellings B C (occasionally not present, but this has not been seen in UK greenhouses)
- 5 pairs of spermathecae

Main characters

Clitellum: 14-16

Female pore: 14

Male pore: 18

Prostates: present D

Intestinal caeca: present D

Spermathecae: 5 pairs: 4/5 5/6 6/7 7/8 8/9

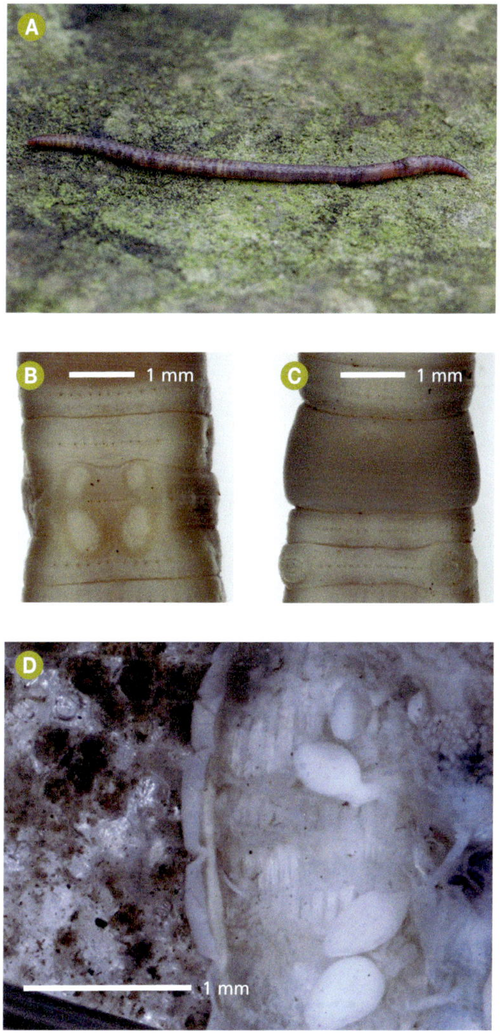

Polypheretima taprobanae
(Beddard, 1892)

Distinctive characters

- No intestinal caeca

Main characters

Clitellum: 14-16

Female pore: 14

Male pore: 18

Prostates: present

Intestinal caeca: absent

Spermathecae: 1 large pair: 8

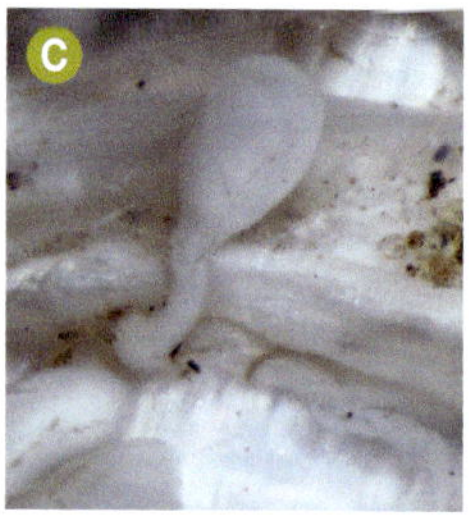

Pontoscolex corethurus
(Müller, 1856)

Very common in glasshouses and artificial environments (as well as around the world)

Main characters

Clitellum: (15)16-22 (23)

Spermathecae: 2 pairs

Intestinal caecas: absent

Long extendable prostomium, so often called the elephant tongue worm)

References and further reading

ANDERSON, F., WILLIAMS, B., HORN, K., ERSÉUS, C., HALANYCH, K., SANTOS, S. and JAMES, S. (2017) Phylogenomic analyses of Crassiclitellata support major northern and southern hemisphere clades and a Pangaean origin for earthworms. *BMC Evolutionary Biology* **17**: 123.

ANDRIUZZI, W., SCHMIDT, O., BRUSSAARD, L., FABER, J.H. and BOLGER,T. (2016) Earthworm functional traits and interspecific interactions affect plant nitrogen acquisition and primary production. *Applied Soil Ecology* **104**: 148-156.

BEDDARD, F.E. (1890) Exhibition of and remarks upon, some living specimens of oriental earthworms, found in a greenhouse in Scotland. *Proceedings of the Zoological Society London* **1890**: 94.

BEDDARD, F.E. (1891) On the structure of two new genera of earthworms belonging to the Eudrilidae and remarks on *Nematodrilus*. *Quarterly Journal of Microscopal Science* **32**: 235-278.

BEDDARD, F.E. (1892) On some species of the genus *Perichaeta* (sensu stricta). *Proceedings of the Zoological Society of London* **1892**: 153-172.

BEDDARD, F.E. (1893) On some new species of earthworms from various parts of the world. *Proceedings of the Zoological Society of London* **1892**: 666-706.

BEDDARD, F.E. (1901) Contributions to the knowledge and systematic arrangement of earthworms. *Proceedings of the Zoological Society of London* **1901**: 187-206.

BEDDARD, F.E. (1906) Annelida, Oligochaeta. *Kew Bulletin Additional Series* **5**: 66-68.

BLAKEMORE, R.J. (2005) British and Irish Earthworms: a checklist of species.
Available at www.annelida.net/earthworms/Britain & Ireland.pdf [accessed 16 February 2018].

BLAKEMORE, R.J. (2008) Seeking consensus on main categories of ecological strategies of earthworms. Advances in Earthworm Taxonomy III. Proceedings of the 3rd International Oligochaete taxonomy meeting. Nicosia, Cyprus 2-6 April 2007, 131-136.

BLAKEMORE, R.J. (2010) Cosmpolitan earthworms: an eco-taxonomic guide to the peregrine species of the world. Verm Ecology, Kippax, ACT, Australia.

BLAKEMORE, R.J. (2012) Hibernian reports of a new Franco-Iberian worm (Oligochaeta: Megadrilacea: Lumbricidae) Opuscula Zoologica **43**(2): 121-130.

BOUCHÉ, M.B. (1971) *Relation entre les structures spatiales et fonctionnelles des écosystèmes illustréspar le rôle pédobiologique des vers de terre*. In: PESSON, P. (ed.) *La vie des sols*. Paris: Gauthier-Villars, 187-209.

BOUCHÉ, M.B. (1972) *Lombriciens de France. Ecologie et systematique*. Paris: INRA.

BOUCHÉ, M.B. (1977) Stratégies lombriciennes. In: LOHM, U. and PERSSON, T. (eds) *Soil organisms as components of ecosystems*. Stockholm: *Ecology Bulletin/NFR* **25**: 122-132.

BRIONES, M.J., OSTLE, T. and PIEARCE, G. (2008) Stable isotopes reveal that the calciferous gland of earthworms is a CO_2-fixing organ. *Soil Biology and Biochemistry* **40**: 554-557.

BRUSSAARD, L., DE RUITER, P.C. and BROWN, G.G. (2007) Soil biodiversity for agricultural sustainability. *Agriculture Ecosystems & Environment*, **121**: 233-244.

BUTT, K. R. and GRIGOROPOULOU, N. (2010) Basic research tools for earthworm ecology. *Applied and Environmental Soil Science*, **2010**: 1-12.

CARPENTER, D., HODSON, M., EGGLETON, P. and KIRK, C. (2007) Earthworm induced mineral weathering: Preliminary results. *European Journal of Soil Biology* **4**: 176-183.

CARPENTER, D., SHERLOCK, E., JONES D.T., CHIMINOIDES, J., WRITER, T., NIELSON, A., BOAG, B., KEITH, A.M. and EGGLETON, P. (2011) Mapping of earthworm distribution for the British Isles and Eire highlights the underrecording of an ecologically important group. *Biodiversity and Conservation* **21**: 475-485.

CERNOSVITOV, L. and EVANS, A.C. (1947) Lumbricidae (Annelida) with a key to the common species. *Synopses of the British Fauna* **6**. London: Linnean Society of London.

COSÍN, D.J.D., NOVO, M. and FERNANDEZ, R. (2011) Reproduction of earthworms: sexual selection and parthenogenesis. In: KARACA, A. (ed.) *Biology of earthworms*. Soil Biology **24**. Berlin: Springer.

CSUZDI, C. (1988) Uber die Bestimmung der Konsummenge von *Eisenia foetida* / Sav. 1826 / Oligochaeta: Lumbricidae. *Opuscular Zoologica* **23**: 137-139.

CSUZDI, C. and PAVLICEK, T. (2002) *Murchieona miniscula*, a newly recorded earthworm from Israel and distribution of the genera *Dendrobaena* and *Bimastos* in Israel. *Zoology in the Middle East* **25**:105-114.

CSUZDI, C. and ZICSI, A. (2003) *Earthworms of Hungary (Annelida: Oligochaeta; Lumbricidae).* Budapest: Hungarian Natural History Museum.

CSUZDI, C., SZLAVECZ, K. (2003) *Lumbricus friendi* Cognetti 1904, a new exotic earthworm in North America. *Northeastern Naturalist* **10**(1): 77-82.

CSUZDI, C., CHANG, C.-H., PAVLICEK, T., SZEDERJESI, T. and SZLAVECZ, K. (2017) Molecular phylogeny and systematics of native North American Lumbricid earthworms (Clitellata: Megadrili). *PLOS ONE* **12**(8): e0181504.

CUENDET, G. (1983) Predation on earthworms by the Black-headed gulls (*Larus ridibundus* L.). In: SATCHELL, J.E. (eds) *Earthworn ecology*. Dordrecht: Springer.

DARWIN, C. (1881) *The formation of vegetable mould through the action of earthworms*. London: John Murray.

DE VRIES F.T., LIIRI M.E., BJØRNLUND, L., BOWKER M.A., CHRISTENSEN, S., SETÄLÄ H.M. and BARDGETT, R.D. (2012) Land use alters the resistance and resilience of soil food webs to drought. *Nature Climate Change* **2**: 276-280.

EDWARDS, C.A. (ed.) (1998) *Earthworm ecology*. New York: St Lucie Press.

EDWARDS, C.A. (ed.) (2004) *Earthworm Ecology* (2nd ed.). Boca Raton: CRC Press LLC.

EDWARDS, C.A. and BOHLEN, P.J. (1996) *Biology and ecology of earthworms* (3rd Ed.) London: Chapman and Hall.

EDWARDS, C.A. and LOFTY, J.R. (1977) *Biology of earthworms* (2nd ed.). London: Chapman and Hill.

FRIEND, H. (1891) Earthworms of the North of England. *The Naturalist, Hull* **16**: 13-15.

FRIEND, H. (1911) New British enchytraeids. *Journal of the Royal Microscopical Society* **31**: 730-736.

FRIEND, H. (1916) Alien Oligochaets in England. *Journal of the Royal Microscopical Society* **36**: 262-271.

FRIEND, H. (1923) *British earthworms and how to identify them*. London: Epworth Press.

FRIEND, H. (1924) *The story of British annelids*. London: Epworth Press.

GERARD, B.M. (1964) British Lumbricidae. *Synopses of the British Fauna* **6**. London: Linnean Society of London.

GRIGOROPOULOU, N., BUTT, K.R. and LOWE, C.N. (2009) Interactions of juvenile *Lumbricus terrestris* with adults and their burrow systems in a two dimensional microcosm. *Pesquisa Agropecuária Brasileira* **44**: 964-968.

GRIGOROPOULOU, N. and BUTT, K.R. (2010) Field investigations of *Lumbricus terrestris* spatial distribution and dispersal through monitoring of manipulated, enclosed plots. *Soil Biology Biochemistry* **42**: 40-47.

GUILD, W.J.McL. (1955) Earthworms and soil structure. In: KEVAN, D.K.McE. *Soil zoology* London: Butterworth, 83-98.

GUNN, A. (1992) The use of mustard to estimate earthworm populations. *Pedobiologia* **36**: 65-67.

HERMAVAN, B. and CAMERON, K.C. (1993) Structural changes in silt loam under long-term conventional or minimum tillage. *Soil Tillage Research* **26**: 139-150.

JONES, D.T. and EGGLETON, P. (2014) *Earthworms in England: distribution, abundance and habitats*. Natural England Commissioned Report NERC145.

KRUUK, H. (1978) Foraging and spatial organisation of the European Badger *Meles meles. Behavioural Ecology and Sociobiology* **4**(1): 75-89.

LAVELLE, P., BIGNELL, D., LEPAGE, M., WOLTERS, V., ROGER, P., INESON, P., HEAL, O.W. and DHILLON, S. (1997) Soil function in a changing world: the role of invertebrate ecosystem engineers. *European Journal of Soil Biology* 33: 159-193.

LEE, K.E. (1959) The earthworm fauna of New Zealand. *New Zealand Department of Scientific and Industrial Research Bulletin* **130**.

LEE, K.E. (1985) *Earthworms: their ecology and relationships with soils and land use*. Sydney: Academic Press.

LOWE, C.N. and BUTT, K.R. (2005) Culture techniques for soil dwelling earthworms: a review. *Pedobiologia* **49**: 401-413.

LOWE, C.N. and BUTT, K.R. (2008) *Allolobophora chlorotica* (Savigny, 1826): evidence for classification as two separate species. *Pedobiologia* **52**(2): 81-84.

LOWE, C.N. and BUTT, K.R. (2008) Preliminary evidence for the adoption of *Allolobophora virescens* (Savigny, 1826). Advances in Earthworm Taxonomy III. Proceedings of the 3rd International Oligochaete taxonomy meeting. Nicosia, Cyprus 2-6 April 2007, 63-139.

MACDONALD, D.W. (1983) Predation on earthworms by terrestrial vertebrates. In: SATCHELL, J.E. (ed.) *Earthworm ecology from Darwin to vermiculture*. London: Chapman & Hall, 393-414.

MELLANBY, K. (1966) Mole activity in woodlands, fens and other habitats. *Journal of Zoology* **149**: 35-41.

MULDOWNEY, J. and SCHMIDT, O. (2002) *Allolobophora cupulifera* in Ireland: First records for the British Isles. *Megadrilogica* **9**(5): 29-32.

NUUTINEN, V. and BUTT, K.R. (2003) Interaction of *Lumbricus terrestris* L. burrows with field subdrains. 7th international symposium on earthworm ecology, Cardiff, 2002.

OPAL (2011) Data quality assessment report: soil and earthworm survey. Available at www.imperial.ac.uk/opal-soil/outputs [accessed 16 February 2018].

POP, V.V. (1998) Earthworm biology and ecology – a case study: the genus *Octodrilus* Omodeo, 1956 (Oligochaeta, Lumbricidae), from the Carpathians. In: EDWARDS, C.A. (ed.) *Earthworm ecology*. Boca Raton: St Lucie Press, 65-100.

SATCHELL, J.E. (1967) Colour dimorphism in *Allolobophora chlorotica* Sav. (Lumbricidae) *Journal of Animal Ecology* **36**: 623-630.

SHERLOCK, E. and CARPENTER, D. (2009) An updated earthworm list for the British Isles and two new 'exotic' species to Britain from Kew Gardens. *European Journal of Soil Biology* **45**: 431-435.

SHERLOCK, E. and JONES, D. (2008) A field guide to British earthworms to support a national survey. Advances in Earthworm Taxonomy III. Proceedings of the 3rd International Oligochaete taxonomy meeting. Nicosia, Cyprus 2-6 April 2007, 225-228.

SIMS, R.W. and GERARD, B.M. (1985) Earthworms. *Synopses of the British Fauna (New Series)* **39**. London: Linnean Society of London.

SIMS, R.W. and GERARD, B.M. (1999) Earthworms. *Synopses of the British Fauna (New Series)* **39**. London: Linnean Society of London.

WANG, C. (2005) Research of ecology immunity in the *Eisenia fetida*: antibiotic barrier system and mechanism of joint presentation. PhD thesis, China Agricultural University.

ZICSI, A. (1974) *A struktúra és funkció Kapcsolata terresztrikus ökoszisztémák földigilisztának tevekénysege tükreben. Doktori ertekezes*, Budapest.

ZICSI, A. (1985) Welche Lumbriciden-Arten eignen sich noch in Europa zum Anlegen von Wurmkulturen zwecks Kompostierungsversuche. *Opuscula Zoologica Budapest*, 21: 137-139.

Appendix

This section contains useful survey forms to help you record earthworms in the field. Forms are also available for free download from the Earthworm Society of Britain website (www.earthwormsoc.org.uk).

The Earthworm Site Survey Form can be used in the field to collect all the information required when undertaking soil pit sampling, mustard extraction and/or microhabitat searches of a site.

The Earthworm Identification Features Sheet (developed by Martin Noble) can be used to conveniently record the salient identification features of an earthworm before working through the key.

Earthworm Site Survey Form

For use when undertaking the National Earthworm Recording Scheme 5-Pit Protocol or a full Earthworm Site Survey.

Site Name:	**Date (dd/mm/yyyy):**
Grid reference	**Recorder Name(s):**
Vice county:	**Habitat:**

Soil Pits

Soil pit sampling undertaken?

Yes ☐ No ☐

Number of soil pits:

No. of juveniles returned to soil:

Mustard Sampling

Mustard sampling undertaken?

Yes ☐ No ☐

Number of mustard sampling points:

No. of juveniles returned to soil:

Microhabitat Searches

Microhabitat searches undertaken?

Yes ☐ No ☐

Microhabitat 1 details:

Microhabitat 2 details:

Microhabitat 3 details:

Comments:

Earthworm Identification Features Sheet

For use when identifying earthworms only: Please submit records using the ESB earthworm records submission sheet or the iRecord Earthworm Survey Form

Date	Site details	Head	Male Pore	Clitellum	TP		Clitellum	Setae	Comments (including TP shape)	Species
		Epi or Tany		Start	Start	End	End	Close or Wide		

Earthworm Site Survey Form

For use when undertaking the National Earthworm Recording Scheme 5-Pit Protocol or a full Earthworm Site Survey.

Site Name:	**Date (dd/mm/yyyy):**
Grid reference	**Recorder Name(s):**
Vice county:	**Habitat:**

Soil Pits	Mustard Sampling	Microhabitat Searches
Soil pit sampling undertaken?	Mustard sampling undertaken?	Microhabitat searches undertaken?
Yes ☐ No ☐	Yes ☐ No ☐	Yes ☐ No ☐
Number of soil pits:	Number of mustard sampling points:	Microhabitat 1 details:
No. of juveniles returned to soil:	No. of juveniles returned to soil:	Microhabitat 2 details:
Comments:		Microhabitat 3 details:

Earthworm Identification Features Sheet

For use when identifying earthworms only: Please submit records using the ESB earthworm records submission sheet or the iRecord Earthworm Survey Form

Date	Site details	Head	Male Pore	Clitellum	TP		Clitellum	Setae	Comments (including TP shape)	Species
		Epi or Tany		Start	Start	End	End	Close or Wide		

Martin Noble

Notes